Sanctuary Not Certain

American, British, Australian, and Canadian Hospital Ships in the European–African–Middle Eastern Theatre in World War II

Cdr. David D. Bruhn, USN (Retired)

During World War II, German and Italian aircraft bombed and sank thirteen Allied hospital ships (7 British, 5 Greek, 1 Norwegian) in the European Theatre. A fourteenth hospital ship (British) was sunk by a German sea mine during the Normandy invasion. As a result of these illegal, purposeful attacks on clearly marked hospital ships, wounded servicemen could not be assured of sanctuary on board presumed safe havens; neither could the ships' crews nor the doctors, nurses and other medical staff on board to provide patient care.

The chain of medical care began with servicemen brought to field hospitals near the front lines, then, if necessary, transferred to more distant evacuation hospitals. Critically wounded patients were further transported to station and general hospitals via hospital ships, hospital trains, or aircraft, attended by nurses and other medical personnel.

Typically, though scarcely recognized by authorities paying tribute to WWII veterans' groups, Allied nurses were on the front lines as well as on board hospital ships in combat zones and at more rearward hospitals. During the invasion of North Africa marking America's entry into the war in Europe, U.S. Army nurses waded ashore from landing craft with assault troops; it being too dangerous for hospital ships to enter port. Over the course of the war, women of the U.S. Army Nurse Corps suffered death and wounds as they treated soldiers, sailors, and airmen.

U.S. Army nurses (in all theatres of war) were decorated for meritorious service and bravery under fire. Decorations included the Distinguished Service Medal, Silver Star, Legion of Merit, Distinguished Flying Cross, Soldier's Medal, Bronze Star Medal, Purple Heart, Air Medal, and Army Commendation Medal.

British, Canadian, Australian, and New Zealand Nursing Sisters serving aboard Commonwealth hospital ships shared these same dangers and hardships of war. But those in charge were also mindful of appearances. A Matron directed her junior Nursing Sisters, "here borrow my comb, and try not to look like survivors!" following several hours in Mediterranean water, clinging to wreckage after their British Army hospital ship was sunk, and having been brought ashore in Oran where their bedraggled uniforms were exchanged for men's clothing.

DEDICATION:

To my wife, former LCDR Nancy P. Bruhn, NC, USNR,
and all military nurses, past, present, and future.

Five reasons why you should marry a nurse:
- Nurses are compassionate
- They know how to handle injuries
- They are good listeners
- They are incredibly patient
- They can create order out of chaos[1]

LTJG Nancy P. Bruhn, NC, USNR, 13 June 1985.

Sanctuary Not Certain

American, British, Australian, and Canadian Hospital
Ships in the European–African–Middle Eastern
Theatre in World War II

Cdr. David D. Bruhn, USN (Retired)

HERITAGE BOOKS
2025

HERITAGE BOOKS

AN IMPRINT OF HERITAGE BOOKS, INC.

Books, CDs, and more—Worldwide

For our listing of thousands of titles see our website
at
www.HeritageBooks.com

Published 2025 by
HERITAGE BOOKS, INC.
Publishing Division
5810 Ruatan Street
Berwyn Heights, Md. 20740

International Standard Book Number
Paperbound: 978-0-7884-5039-6

Heritage Books by Cdr. David D. Bruhn, USN (Retired)

Battle Stars for the "Cactus Navy":
America's Fishing Vessels and Yachts in World War II

Beavers: American River College's Running Dynasty, 1964–1979
David D. Bruhn and Al Baeta

Distant Finish
David C. Bruhn and Jack Leydig

Enemy Waters:
Royal Navy, Royal Canadian Navy, Royal Norwegian Navy,
U.S. Navy, and Other Allied Mine Forces Battling the
Germans and Italians in World War II
Cdr. David D. Bruhn, USN (Retired)
and Lt. Cdr. Rob Hoole, RN (Retired)

Eyes of the Fleet:
The U.S. Navy's Seaplane Tenders and
Patrol Aircraft in World War II

Gators Offshore and Upriver:
The U.S. Navy's Amphibious Ships and Underwater Demolition Teams,
and Royal Australian Navy Clearance Divers in Vietnam

Guns Up, Depth Charges Readied:
U.S. Navy, Commonwealth, and Other Allied Escort Ships
Shepherding Convoys, and Battling German and Italian Air
and Naval Forces in the Mediterranean in World War II

Guns Up:
Naval Action in the Yellow Sea off Korea, 1950–1953

Home Waters:
Royal Navy, Royal Canadian Navy, and U.S. Navy
Mine Forces Battling U-Boats in World War I
Cdr. David D. Bruhn, USN (Retired)
and Lt. Cdr. Rob Hoole, RN (Retired)

Ingram's Fourth Fleet:
U.S. and Royal Navy Operations Against German Runners, Raiders,
and Submarines in the South Atlantic in World War II

Intercept:
The U.S. Navy's Intelligence-Gathering Ships
("Cold War Spy Fleet") 1961–1969, 1985–1989

Kissing Cousins:
U.S. Navy Wooden Minesweepers and Variants (YMS, PCS, AGS)
and USN and Royal Australian Navy Bomb and Mine Disposal
Personnel in the Pacific in World War II, 1944–1945

Land Yacht Seaward*:*
Building a Cozy Wooden Camper for a Small Truck

MacArthur and Halsey's "Pacific Island Hoppers":
The Forgotten Fleet of World War II

Nightraiders:
U.S. Navy, Royal Navy, Royal Australian Navy, and
Royal Netherlands Navy Mine Forces Battling
the Japanese in the Pacific in World War II
Cdr. David D. Bruhn, USN (Retired)
and Lt. Cdr. Rob Hoole, RN (Retired)

On the Gunline:
U.S. Navy and Royal Australian Navy Warships off Vietnam, 1965–1973
Cdr. David D. Bruhn, USN (Retired)
and STGCS Richard S. Mathews, USN (Retired)

Queenstown Bound:
U.S. Navy Destroyers Combating German U-boats
in European Waters in World War I

Rarely Idle:
U.S. Navy Sub-chasers and Royal Navy Motor Launches
and Canadian-built Drifters Combating
German U-boats in World War I
Cdr. David D. Bruhn, USN (Retired),
Lt. Cdr. Rob Hoole, RN (Retired)
and George H. S. Duddy

Ready to Haul, Ready to Fight:
U.S. Navy, Royal Australian Navy, and British Merchant
Navy Cargo Ships in the Pacific in World War II

Salvation from the Sky:
U.S. Navy, Royal Australian Air Force, and Royal New Zealand Air
Force Heroic Air-Sea Rescue in the Pacific in World War II
Cdr. David D. Bruhn, USN (Retired)
and Stephen Ekholm

Sanctuary Not Certain:
American, British, Australian, and Canadian Hospital Ships in the
European–African–Middle Eastern Theatre in World War II

Sandscrapers:
The U.S. Navy's LSMs (Medium Landing Ships)
and LSM(R)s (Rocket Ships) in World War II

Send Some King's Ships:
U.S. Navy, Royal Naval Patrol Service, and Royal Canadian Navy Ships
Combating German U-boats off North America's Eastern Seaboard,
and RNPS and South African Naval Forces Vessels in
African Waters as well, 1942–1945
Cdr. David D. Bruhn, USN (Retired)
and Lt. Cdr. Rob Hoole, RN (Retired)

Stand Easy:
Creating a Small British Pub, and Considerable Comradeship,
in the Corner of a Garage

Stream Gear:
U.S. Navy, Royal Navy, Royal Australian Navy, South African
Naval Forces, and Royal Hellenic Navy Minesweepers'
Dangerous Operations in the Mediterranean in World War II.
Cdr. David D. Bruhn, USN (Retired)
and Lt. Cdr. Rob Hoole, RN (Retired)

Stride Out

Support for the Fleet:
U.S. Navy and Royal Australian Navy Service
Force Ships That Served in Vietnam, 1965–1973

Toe the Mark

Turn into the Wind:
Volume I: US Navy and Royal Navy Light Fleet Aircraft Carriers
in World War II, and Contributions of the British Pacific Fleet

Turn into the Wind:
Volume II: US Navy, Royal Navy, Royal Australian Navy, and
Royal Canadian Navy Light Fleet Aircraft Carriers in the
Korean War and through End of Service, 1950–1982

War Bound from Stockton: U.S. Navy Ships from California's
Central Valley, In Harm's Way in Pacific Waters in World War II

We Are Sinking, Send Help!:
The U.S. Navy's Tugs and Salvage Ships in the African,
European, and Mediterranean Theaters in World War II

Wooden Ships and Iron Men:
The U.S. Navy's Coastal and Motor Minesweepers, 1941–1953

Wooden Ships and Iron Men:
The U.S. Navy's Coastal and Inshore Minesweepers, and
the Minecraft that Served in Vietnam, 1953–1976

Wooden Ships and Iron Men:
The U.S. Navy's Ocean Minesweepers, 1953–1994

Contents

Photos and Illustrations

Maps and Diagrams

Foreword

In his latest book, *Sanctuary Not Certain*, David Bruhn provides a valuable insight into a generally little-known subject, the Allied hospital ships in the European Theatre in World War II. The British interacted with the few Australian hospital ships in the Mediterranean in the early years of the war and this foreword provides a summary of the Australian hospital ships.

During the first years of World War II, Australia's military strategy was closely aligned with that of the United Kingdom's and consequently most Australian military units that were deployed overseas in 1940 and 1941 were sent to the Mediterranean and Middle East where they formed an integral part of the Commonwealth forces in the area. The three Australian Imperial Force (AIF) infantry divisions despatched to the Middle East were subsequently heavily involved in the fighting that followed.

Following the outbreak of war in the Pacific in December 1941 most elements of the AIF including the 6th and 7th Divisions, as well as the Royal Australian Navy (RAN) ships, returned to Australia in early 1942 to counter the perceived Japanese threat to Australia. The 9th Division remained to play an important role in the Allied victory at El Alamein in October 1942 before it also left for the Pacific in January 1943.

During World War II, Australia's armed forces were supported by four fully equipped hospital ships: *Manunda* and *Wanganella*, converted from relatively new and well-found passenger liners; *Oranje*, a palatial liner converted, run and maintained by the Dutch Government; and, in the Pacific War, *Centaur*, a British registered motor ship. Australian-based hospital ships were operated by the Australian Shipping Control Board (ASCB) and were known as HMAHS (His Majesty's Australian Hospital Ship).

Manunda, Wanganella and *Centaur* were taken over by arrangement between the Australian Government and the owners for conversion as hospital ships, and the alterations were carried out under the direction of the RAN. The method of control adopted on Australian hospital ships was combined, under which the RAN supervised or carried out the maintenance and routing of hospital ships, while the merchant navy sailed the ships, and the army controlled the hospital as a military unit.

In February 1941 the Netherlands East Indies government offered the use of the *Oranje*, a fast new motor liner of 20,000 tons hitherto reserved for service as an auxiliary cruiser, as a hospital ship on the Middle East – Australia – New Zealand run. The Dutch authorities offered to equip, man and operate the ship at their own expense; this most generous offer was accepted by Australia and the ship was converted at Cockatoo Island dockyard in Sydney. Although she was under Australian command, she kept her Dutch crew and remained under the Dutch flag throughout the war.

Manunda and *Wanganella*, whose size (9,000 to 10,000 tons gross) and lay-out were similar, proved very satisfactory as hospital ships carrying a little under 400 patients, and 150 more in canvas cots in an emergency. *Manunda* was a motorship of Australian registry built in 1929. She was requisitioned as a hospital ship in July 1940 and between November 1940 and September 1941 completed four round trips to the Middle East.

Photo Foreword-1

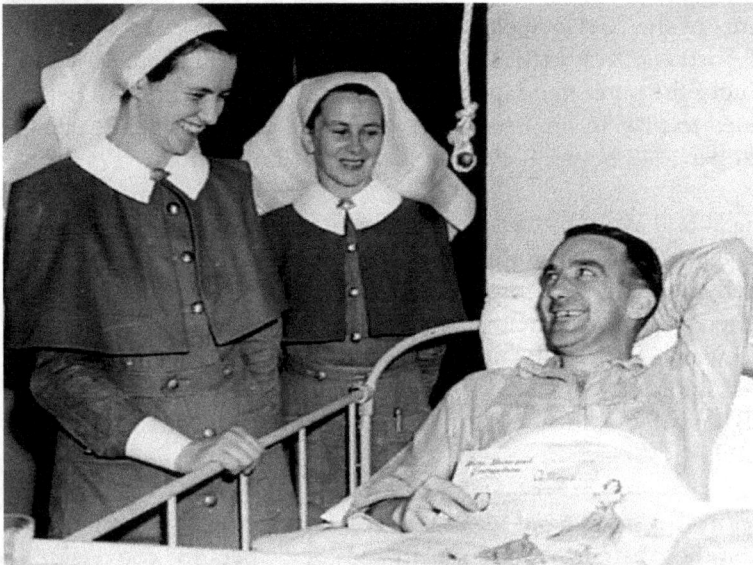

Arrival of Hospital Ship *Manunda* with her first patients at Melbourne in December 1940.
Australian War Memorial photograph 004314

The possible dangers run by a hospital ship were illustrated during *Manunda*'s second voyage, when she was delayed for some weeks at Ismailia, Egypt on the west bank of the Suez Canal by the mining of the Canal by enemy aircraft. While in the Great Bitter Lake (a part of the Suez Canal), a Greek ship transiting in advance of the *Manunda* struck a mine, and *Manunda* ran further risks by being tied up in the vicinity of a vessel with a load of explosives while air raids were going on.

Map Foreword-1

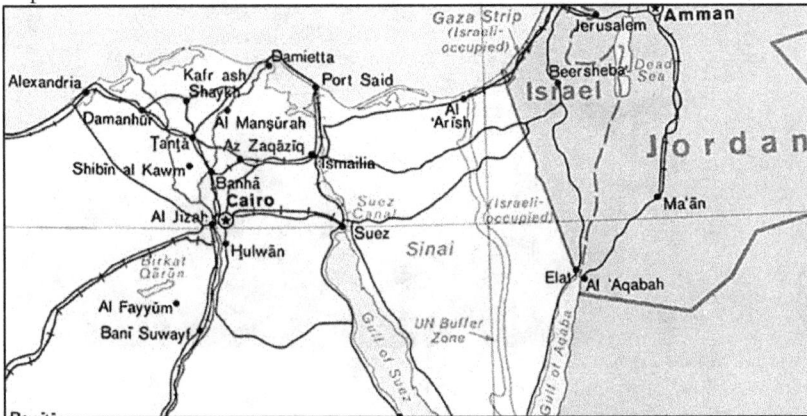

Northeast Egypt

Wanganella was also an Australian motorship, and her conversion to a hospital ship was completed in July 1941. Her first voyage as HMAHS was to Singapore carrying the 2/13th Australian General Hospital, returning late September to Sydney. The next voyage was to Suez to collect wounded for return to Sydney and Brisbane. In November 1941, whilst at Port Tewfik at the entrance to the Suez Canal, she experienced a bombing raid, but came through without damage. After two more voyages to the Middle East the ship was ordered to Port Moresby in May 1942 to collect injured for return to Australia, and to transport US casualties from Townsville to the US 4th General Hospital which was temporarily housed in the newly constructed replacement for the Royal Melbourne Hospital. In the earlier part of the Pacific war the *Wanganella* like the *Manunda* served the New Guinea ports and visited Milne Bay and later still went to the Solomons and to Borneo.

Photo Foreword-2

Hospital Ship *Wanganella* departing Port Tewfik, Suez,
with a shipload of wounded, November 1941.
Australian War Memorial photograph 021225

During the period 1940-42 Australian sick and wounded were
transported over both short and long sea routes. To and from Australia
and the Middle East the distances were long, but risks were slight,
though some uncertainty was felt about whether raiders might interfere
with protected ships. These trips were made by hospital ships or
returning transports according to the needs of the sick or unfit, and
contrasted with the short but hazardous journeys along the North
African coast and to and from Greece and Crete. For short runs in the
Mediterranean British hospital ships were often used for Australian
casualties, there being no Australian hospital ships on this run. In
addition, defensively armed ships were employed, and other transports
as available.

Oranje completed two runs to the Middle East and back before
December 1941 when the Dutch declared war on Japan, and in these
trips her speed and capacity (650 beds) enabled her to transport 900
men in short time in a high degree of comfort. It is interesting in the
light of hindsight that on nearing Australian waters during her second
trip, she ran close to the position where in November 1941, the light

cruiser HMAS *Sydney* encountered the German auxiliary cruiser *Kormoran* and both were lost in the resultant action.

Originally staffed and operated by a Dutch crew, with a small complement of Australian and New Zealand staff, *Oranje* later carried a largely Australian, and eventually a largely New Zealand medical staff. She completed 41 war voyages, covering over 382,000 nautical miles and carrying some 32,461 patients. After the Australian troops had been withdrawn from the Middle East at the end of 1942, the war took a different shape for Australia, and with the focus on the South-West Pacific and the great need for medical officers in Australia, all Australians were taken off the *Oranje* and replaced by New Zealanders. Under New Zealand's command, the *Oranje* played a crucial role in transporting and caring for sick and wounded Allied patients. The *Oranje* was a most valuable addition to the fleet of hospital ships serving the Middle East, and after the end of the war this ship again rendered service to Australia by transporting prisoners of war.

Photo Foreword-3

Patients boarding the hospital ship *Oranje* in Port Tewfik, Suez, circa August 1941. Australian War Memorial photograph 020068

Early in 1943 the Australian Army took over the *Centaur* for conversion to a hospital ship. At first this ship of 3,200 tons was planned as a hospital carrier fitted for short trips, but so many improvements were incorporated that eventually *Centaur* was converted to a modern hospital ship capable of carrying 280 cot cases on voyages of up to 18 days, if required.

Centaur departed from Sydney on 12 May 1943 on her first operational deployment as a hospital ship, with no patients but with crew, medical staff and stores for the 2/12th Field Ambulance, having a total of 332 persons aboard. About 0400 on 14 May she was torpedoed northeast of Brisbane by the Japanese submarine *I-177*. *Centaur* was painted in hospital ship colours and was fully illuminated at the time. There were just 64 survivors. The wreck was identified on 20 December 2009 at a depth of 2,000 metres, and the site is now gazetted as a war grave.

The New Zealand Second Division remained with the British Eighth Army in North Africa and Italy but the NZ Hospital Ship *Maunganui* alone, however, was not sufficient to bring back to New Zealand all the invalids from the Middle East. The Hospital Ships *Wanganella* and *Oranje* assisted on occasions, ensuring the sick and wounded were evacuated throughout the War.

The most outstanding events of the period 1943-45 in the Pacific as far as hospital ships were concerned, were the commissioning and loss of the *Centaur*, the changes caused by the development of amphibious warfare and combined landings, and an increase in the air transport of sick and wounded.

Commodore Hector Donohue AM RAN (Rtd)

Foreword

It wasn't until David enlisted my help in researching his enlightening book that I realised the significant role played by hospital ships in various conflicts and the tragedies associated with so many of them; eight out of 42 British hospital ships were lost to enemy action in the Second World War. As a former Royal Navy officer, I took a particular interest in the vessels requisitioned for use by the Royal Navy and British Army as hospital ships.

Hospital ships tended to be merchant vessels taken up from trade while continuing to be operated by Merchant Navy personnel. The Royal Navy and the British Army provided the embarked medical staff and supporting personnel. Almost without exception, hospital ships and hospital carriers were given the prefix HMHS and wore the blue ensign defaced by a horizontal anchor indicating a government service ship (flags are 'worn' by ships but 'flown' by high-ranking personages on board them) whether they operated under the auspices of the Admiralty for the Royal Navy or the War Department for the British Army. It wasn't until 1964 that the Royal Navy, British Army and Royal Air Force merged under a central Ministry of Defence.

To determine which hospital ships were administered by the Royal Navy during the Second World War, I consulted the Navy List archives which list these vessels at some time or other during the period. Significantly, the hospital ship *Maine*, which carried the preface HMHS during her previous service in the Great War (WWI), was the Royal Fleet Auxiliary RFA *Maine* during her WWII hospital ship service.

- HMHS *Aba* – based at Devonport, Plymouth
- HMHS *Amarapoora* – based at Devonport, Plymouth
- HMHS *Cap St Jacques*
- HMHS *Empire Clyde* (formerly *Leonardo da Vinci*)
- HMHS *Gerusalemme*
- HMHS *Isle of Jersey* – based at Portsmouth
- RFA *Maine* (formerly *Panama*) – based at Chatham
- HMHS *Ophir*
- HMHS *Oxfordshire* – based at Chatham
- HMHS *Tjitjalengaka*
- HMHS *Vasna* – based at Portsmouth
- HMHS *Vita*

The only personnel mentioned for these ships are the naval medical staff, chaplain and paymaster embarked. Each of the services had its own medical corps, as they have today. Certain other hospital ships were manned by Army medical personnel, just as those in *The Navy Lists* were manned by RN surgeons (e.g. Surg Lt RN or RNVR), dentists (e.g. Surg Lt[D] RN or RNVR), wardmasters (male equivalent nurses, e.g. Wt [Warrant] Wardmaster), and reserve nurses (all female in those days) who had their own officer and rating (enlisted) titles and ranking system.

The medical staff embarked in Army hospital ships and hospital carriers comprised doctors, orderlies and other supporting staff belonging to the RAMC (Royal Army Medical Corps) and matrons, nursing sisters and staff nurses belonging to the QAIMNS (Queen Alexandra's Imperial Military Nursing Service) and its reserves and/or the TANS (Territorial Army Nursing Service). QAIMNS became a corps in the British Army and was renamed the QARANC (Queen Alexandra's Royal Army Nursing Corps) in 1949. The Territorial Army was the active-duty volunteer reserve force of the British Army.

The medical staff embarked in Naval hospital ships and hospital carriers comprised doctors and dentists, both using the prefix 'Surgeon' in their normal officer rank, of the regular Royal Navy and Royal Naval Volunteer Reserve as well as matrons, various grades of nursing sister, and nurses belonging to the QARNNS (Queen Alexandra's Royal Naval Nursing Service) and its reserve.

Naval and the Army hospital ships also carried chaplains. The British Army has always assigned its chaplains ranks equivalent to their officer counterparts and treated them to time-served promotions. Conversely, the RN has never given its chaplains ranks although they are notionally equivalent to Cdr in status and pay. This is so they can be regarded as equal, in the tradition of "all of one company," by whomever they are giving pastoral care.

Entries from *The Navy List* of December 1939 provide insights on both military and civilian personnel manning on board hospital ships. The Master (ship's captain), Ch. (Chief) Officer, and Ch. (Chief) Engineer assigned to the RFA *Maine* were civilian merchant mariners.

HOSPITAL SHIPS

ABA (*Dev*).

Surg. Capt.	S. Bradbury, MD, DPH (*ret*)	3 Sept 39
Surg. Com. R.N.V.R.	St. G. B. D. Gray, MD, BS, FRCS, LRCP	3 Sept 39
Surg. Lieut.-Com.	C. N. H. Joynt, MB, BCh	3 Sept 39
Surg. Lieut.-Com. R.N.V.R.	P. B. Moroney, MB, ch.B	3 Sept 39
Surg. Lieut., R.N.V.R.	J. M. Robertson, MB, BS	3 Sept 39
	J. M. Ridyard, MSC, MB, chB	3 Sept 39
	C. P. Hay, MD, ChB, DPH, MRCP	3 Sept 39
Chaplain	Rev. F. Freeman, MA (*ret*)	8 Sept 39

AMARAPOORA (*Dev*).

Surg. Capt.	J. H. B. Martin, MD, BCh (*ret*)	3 Sept 39
Surg. Com. R.N.V.R.	W. J. Payne, MB, chB, FRCS	23 Oct 39
Surg. Lieut.-Com.	D. D. Steele-Perkins, LRCP & S	23 Oct 39
Surg. Lieut.-Com. R.N.V.R.	I. O. Clyde, MRCS, LRCP	23 Oct 39
	W. H. Foy, MRCS, LRCP, DMRE	4 Oct 39
	E. A. H. Hindhaugh, FRCS, LRCP, MB, BS	18 Oct 39
Surg.-Lieut.-Com. (D), R.N.V.R.	F. E. D. Hallon	7 Oct 39
Tempy. Surg. Lieut., R.N.V.R.	C. A. Clarke, MD, MRCP	23 Oct 39
Tempy. Chaplain, R.N.V.R.	Rev. W. H. Brown, MA	4 Nov 39

ISLE OF JERSEY (*Po*).

Surg. Capt.	B. R. Bickford, DSO, MRCS, LRCP, DOMS (*ret*)	28 Aug 39
Surg. Com.	P. B. Jackson, MRCS, LRCP	28 Aug 39
Surg. Lieut.-Com. R.N.V.R.	H. G. Ungley, MD, BS, LRCP, FRCS	28 Aug 39
	R. D. Bradshaw, MD, ChB	28 Aug 39
	H. E. Holling, MSC, LMB, chB, MRCP	28 Aug 39
	S. B. Levy, MRCS, LRCP	28 Aug 39
Surg. Lieut., R.N.V.R.	H. B. Howell, MRCS, LRCP	28 Aug 39
Chaplain	Rev. V. R. Bartlett	5 Oct 39

MAINE (*late PANAMA*) (*Ch*).

Surg. Capt.	M. S. Moore, MD, BCh, DPH	15 July 39
Surg. Com.	(O) R. R. Bake, MB, BCh	22 Oct 38
	(S) T. F. Crean, LRCP & S	30 Apr 39
Surg. Com., R.N.V.R.	G. F. Abercrombie, MD, BCh	19 Aug 39
Paym. Com.	G. A. J. Scholfield	28 Aug 37
Surg. Lieut.-Com.	(A) A. L. Moorby, MRCS, LRCP	13 June 38
Surg. Lieut.-Com. (D)	P. S. Turner, LDS	21 Feb 38
Surg. Lieut.	D. C. Dobson, MB, ChB	11 July 39
Surg. Lieut., R.N.V.R.	H. W. Clegg, MB, chB, DPH	19 Aug 39
	H. R. I. Wolfe, MD, BS, FRCS	19 Aug 39
Chaplain R.N.V.R.	Rev. R. A. Deane, MA	5 Sept 39
Wt. Wardmaster	R. E. Dickie	10 Jan 38
Master	W. Whiteley	30 Apr 36
Ch. Officer	D. C. Snowden	17 Feb 39
Ch. Engineer	F. C. Reynolds	10 June 37

OXFORDSHIRE (*Ch*).

Surg. Capt.	H. D. Drennan, DSO, MB, BCh (*ret*) (*Principal Medical Officer*.)	3 Sept 39
Surg. Com., R.N.V.R.	R. Wear, MD, BS, MRCS, LRCP, DMRE	3 Oct 39
Surg.-Com. (D) R.N.V.R.	G. A. O. White, LDS	24 Oct 39
Surg. Com.	J. H. Nicolson, MB, chB	24 Oct 39
Surg. Lieut.-Com. R.N.V.R.	D. W. Bawtree, MB, BCh, FRCS (Ed)	19 Oct 39
	R. F. Woolmer, MD, BCh	19 Oct 39
Surg. Lieut., R.N.V.R.	W. G. Campbell, LDS, MB, chB, FRCS	3 Oct 39
	E. M. Buzzard, MRCS, LRCP, BM, BCh	19 Oct 39
Chaplain, R.N.V.R.	Rev. J. S. Billings, BA	19 Oct 39

VASNA (*Po*).

Surg. Capt.	F. G. Hunt, MRCS, LRCP (*ret*)	19 Aug 39
Surg. Lieut.-Com.	G. S. Thoms, MB, chB	26 Aug 39
Surg. Lieut.-Com. R.N.V.R.	G. I. Puddy, MBBS, MRCP, LRCP	28 Aug 39
	S. C. Suggit, MBBS, FRCS, LRCP	28 Aug 39
Surg. Lieut.-Com. (D), R.N.V.R.	A. Brook Bateman, MD, chB, BDS	23 Sept 39
Surg. Lieut., R.N.V.R.	H. J. Wade, MB, BSc	26 Aug 39
	F. J. S. Gowar, MBBS, FRCS, LRCP	26 Aug 39
	D. R. Maitland, MB, chB, MRCP	26 Aug 39
Chaplain, R.N.V.R.	Rev. H. M. W. Hocking, MA (*pmby*)	5 Sept 39
Wt. Wardmaster	C. H. Thompson (*act*)	3 Nov 39

Entries from *The Navy List* of July 1945 (next two and one-half pages) evidence the increase in the number of Naval hospital ships through the war:

HOSPITAL SHIPS

AMARAPOORA (Dev).

Surg. Capt.	A. B. Clark, MB, BS, DPH (ret)	23 Oct 43
Surg.-Com., R.N.V.R.	(E) J. B. Hutchison, FRCS, LRCP	14 Dec 43
Surg. Lieut.-Com., R.N.V.R.	(S) F. J. S. Gowar, MB, BS FRCS, LRCP	20 July 44
Tempy. Surg. Lieut.-Com., R.N.V.R.	(M) A. C. C. Hughes, MB, BCH, MRCS, MRCP (act); (X) J. K. Armstrong, MB, BCH (act)	1 Sept 44 / 16 Dec 43
Surg. Lieut.-Com. (D)	J. D. Atkins, LDS (emgcy)	9 May 44
Tempy. Surg. Lieut., R.N.V.R.	V. M. Franklin, MB BCH; E. M. Southern, MRCS, LRCP; (A) I. C. W. English, MRCS, LRCP	24 May 44 / 20 July 44 / 11 Aug 44
Tempy. Chaplain, R.N.V.R.	Rev. K. W. Breed, BA	16 Oct 44
Tempy. Lieut. (S), R.N.R.	J. H. Froude	11 June 40
Reserve Senior Sister	Miss E. E. Christian (act)	30 Apr 45
Reserve Nursing Sister	Miss M. Haigh; Miss E. Hazeldene; Miss M. S. Neilson	— Apr 45 / — Apr 45 / — Apr 45
Tempy. Wt. Wardmaster	F. D. Peters	15 May 44

CAP ST. JACQUES.

Surg. Capt.	J. F. Ainley, MB, BCH (ret)	10 Apr 45
Surg. Com.	(M) J. C. Gent, MRCS, LRCP	18 Apr 45
Surg. Lieut.-Com., R.A.N.R.	A. McL. Millar, MB, BS	14 Apr 45
Surg. Lieut.-Com., R.N.V.R.	(S) A. F. M. Barron, MD, FRCS	14 Apr 45
Tempy. Surg. Lieut.-Com., R.N.V.R.	(A) G. Morrice, MB, CHB (act)	16 Apr 44
Tempy. Surg. Lieut., R.N.V.R.	(B) I. MacKenzie, MRCS, LRCP; J. M. Berry, MB, CHB; K. S. Preston, MB, CHB (proby)	14 Apr 45 / 18 Apr 45 / 16 Apr 45
Temp. Lieut.-Com. (S), R.N.V.R.	S. H. Walkem (act)	1 May 45
Reserve Nursing Sister	Miss C. A. Browne; Miss C. J. Dunn; Miss D. M. Press; Miss E. V. Robertson; Miss O. Seed; Miss K. M. Shottle; Miss E. W. Williams	14 Apr 45 (each)
Tempy. Wt. Wardmaster	W. H. P. Wilkinson	7 May 45

EMPIRE CLYDE.

Surg. Capt.	A. de B. Joyce, MB, CHB, DOMS (act)	12 May 45
Surg. Lieut.-Com., R.N.V.R.	(M) T. Colver, MB, CHB, MRCP	18 May 45

GERUSALEMME

Tempy. Surg. Lieut.-Com., R.N.V.R.	(B) O. G. Lloyd, MB, MRCS, LRCP (act)	26 May 45
Tempy. Surg. Lieut.-Com. (D), R.N.V.R.	J. M. Alexander, BDS (act)	26 May 45
Tempy. Surg. Lieut., R.N.V.R.	(S) E. L. Nicolson, MB, CHB; C. J. Briggs, MB, BS (proby); M. J. G. Davies, MB, BS; (A) J. A. Watt, MB, CHB; P. L. Nicolls, MRCS, LRCP	14 May 45 / 22 May 45 / 23 May 45 / 27 May 45 / 1 June 45
Tempy. Chaplain, R.N.V.R.	Rev. T. H. Norman (proby)	28 June 45
Lieut.-Com. (S), R.N.R.	W. E. Sandifer (act)	28 May 45
Matron	Miss J. M. Woolgate (act)	25 May 45
Reserve Nursing Sister	Miss O. M. Codling; Miss M. Cullinans; Miss M. Neeson; Miss I. M. MacLeod; Miss E. Parslow; Miss J. Slater; Miss C. I. Woods	25 May 45 (each)
Tempy. Wt. Wardmaster	S. F. Jerrard	12 May 45

GERUSALEMME

Surg. Capt.	R. W. Higgins, OBE, MB, BCH	14 Feb 45
Surg. Com.	(S) W. V. Beach, FRCS, LRCP	10 Feb 45
Surg. Lieut.-Com.	(B) A. G. G. Toomey, MRCS, LRCP	1 Feb 45
Surg. Lieut.-Com., R.A.N.R.	(G) P. N. Simons, MB, BS	— Feb 45
Surg. Lieut.-Com., R.N.V.R.	(G) R. A. Mogg, MB, BCH, FRCS	31 Jan 45
Tempy. Surg. Lieut.-Com., R.N.V.R.	(A) C. S. Drawner, MRCS, LRCP	3 Feb 45
Tempy. Surg. Lieut., R.N.V.R.	(M) M. D. Young, MB, BCH, MRCP; C. E. Drew, MB, BS	6 Feb 45 / — Feb 45
Tempy. Surg. Lieut. (D), R.N.V.R.	D. Munns, LDS	— Feb 45
Tempy. Chaplain, R.N.V.R.	Rev. G. O. C. Duxbury, MA	— Feb 45
Tempy. Lieut. (S), R.N.V.R.	C. E. Lester	21 Feb 45

ISLE OF JERSEY (Po).

Surg. Com.	A. Long, MRCS, LRCP	7 Apr 45
Surg. Lieut. Com.	(S) T. F. Davies, LMSSA	12 Apr 45
Surg. Lieut.-Com., R.N.V.R.	(X) H. B. Howell, MRCS, LRCP	16 Dec 44

HOSPITAL SHIPS—contd.

Tempy. Surg. Lieut., R.N.V.R.	(A) J. W. Warrick, MRCS, LRCP	21 Mar 45
	(M) J. F. Smith, MB, BCH, MRCP	10 Apr 45
Reserve Nursing Sister	Miss J. Donlan	28 Mar 45
	Miss J. Hallahan	16 Apr 45
	Miss E. L. Pearce	13 Mar 45
Tempy. Wt. Wardmaster	D. C. Jenkins	12 Oct 43

MAINE (Ch). (late PANAMA).

Surg. Com.	(B) J. F. Heggie, MB, ChB	19 Dec 42
Surg. Lieut.-Com.	(O) E. J. Littlodale, LMSSA	30 July 43
	(G) F. B. B. Weston, MRCS, LRCP	26 July 44
Surg. Lieut.-Com., R.N.V.R.	(S) R. W. Smith, MB, ChB.	4 Jan 44
Tempy. Surg. Lieut.-Com., R.N.V.R.	(M) J. R. Forbes, MB, MRCS (act)	18 Apr 45
Surg. Lieut.	D. C. Dobson, MB, ChB	11 July 39
Tempy Surg. Lieut., R.N.V.R.	(A) C. A. Cheatle, MRCS, LRCP	1 May 45
Tempy. Surg. Lieut. (D), R.N.V.R.	J. T. Powell-Cullingford, LDS	4 Jan 45
Tempy. Chaplain, R.N.V.R.	Rev. B. J. Kett, MA	11 Aug 43
Tempy. Lieut.-Com. (S), R.N.V.R.	I. D. T. Clarke (act)	— May 44
Reserve Nursing Sister	Miss D. Gilmore	5 May 45
	Miss J. Jenner-Fust	18 June 45
	Miss M. Kelliher	18 June 45
	Miss D. Lambert	15 May 45
Tempy. Wt. Wardmaster	E. C. Thompson	15 Apr 43

OPHIR

Surg. Capt. R.N.V.R.	F. L. Cassidi, MB, BCH	1 Jan 45
Surg. Com.	(M) A. E. Flannery, LRCP, LRCS	26 Apr 45
Surg. Lieut.-Com., R.N.V.R.	(A) G. A. Hart, MB, BS	18 Oct 44
	(S) J. M. Robertson, MB, BS, MRCS	14 Oct 44
*Tempy. Surg. Lieut.-Com. (D), R.N.V.R.	J. E. Forrest, LDS (act)	20 Jan 45
Tempy. Surg. Lieut., R.N.V.R.	G. A. S. Anthony, MRCS, LRCP	22 July 44
	J. D. Manning, MRCS, LRCP	— Jan 45
Tempy. Chaplain, R.N.V.R.	Rev. R. L. Cole, MA	1 Jan 45
Tempy. Lieut. (S), R.N.V.R.	L. G. Bayley	— Jan 43
Reserve Nursing Sister	Miss W. Allen	10 May 45
	Miss A. M. Gadd	25 May 45
	Miss L. B. Nicholson	19 June 45
	Miss E. D. Russell	23 Apr 45
	Miss J. L. Stow	23 Apr 45
Tempy. Wt. Wardmaster	H. H. Davies	— May 45
	F. L. Timberlake	25 May 45

OXFORDSHIRE (Ch.)

Surg. Lieut.-Com. R.N.V.R.	(H) H. M. Willoughby, VD, MRCS, LRCP, DPH, DTM & H (act)	25 Aug 44
Surg. Lieut.-Com.	(X) E. B. Bradbury, MB, BCH, BAO	3 Feb 44
	(E) J. Dow, MB, ChB	1 Feb 44
	(O) J. Thomas, DSC, BSC, MRCS, LRCP	8 Aug 44
	(M) P. Jones, MB, ChB, MRCP (emgcy)	10 July 44
Surg. Lieut.-Com., R.N.V.R.	(S) P. G. C. Martin, VD, MB, BCH, FRCS	3 Aug 44
Tempy. Surg. Lieut.-Com., R.N.V.R.	(B) J. D. Bradley-Watson, MB, BS, MRCS, LRCP (act)	2 Sept 44
Tempy. Surg. Lieut.-Com. (D), R.N.V.R.	J. T. Cumming, LRCP, LRCS, LRFPS, LDS (act)	8 Oct 43
Tempy Surg. Lieut., R.N.V.R.	G. W. Baker, MB, ChB, FRCS	1 Aug 44
	(A) J. R. Elliott, MRCS, LRCP, DA	5 Aug 44
Tempy. R.C. Chaplain	Rev. G. Pitt	20 July 43
Tempy. Chaplain, R.N.V.R.	Rev. H. Hill, BA	22 Jan 44
Tempy. Lieut. (S), R.N.R.	T. Edwards	29 Oct 42
Reserve Nursing Sister	Miss I. H. Harrower	24 Sept 45
	Miss E. M. Rees	— May 45
Tempy. Wt. Wardmaster	R. E. M. Pryce	9 Sept 44

TJITJALENGKA

Surg. Capt.	F. G. Hunt, MB, BCH	21 Nov 44
Surg.-Com.	(S) W. J. F. Guild, M.D. BCH	19 July 44
Surg. Com., R.N.V.R.	(M) E. G. Brewis, MD, BS, MRCP, DPH (act)	12 Apr 45
Surg. Lieut.-Com.	(O) W. H. C. M. Hamilton, MB, BCH	9 Sept 44
Tempy. Surg. Lieut.-Com.(D), R.N.V.R.	N. K. Davison, LDS (act)	20 Oct 44
Tempy. Surg.-Lieut., R.N.V.R.	(A) R. M. Chalmers, MB, ChB	11 Aug 44
	(E) J. G. Cutts, MRCS, LRCP	— July 44
	(X) G. A. Stevenson, MRCS, LRCP	20 July 44
	G. B. Shaw, MB, ChB	1 Mar 45
	(G) N. S. Craig, MD, DPH	— Oct 44
Chaplain	Rev. G. Bower, BA	30 Nov 44
Matron	Miss M. Kennedy (act)	14 Apr 45
Reserve Nursing Sister	Miss O. Calvert	23 Apr 45
	Miss E. M. Hoad	4 May 45
Tempy. Wt. Wardmaster	F. W. Titmus	26 May 45

3114	THE ROYAL NAVY

VASNA (Dev.)		VITA	
Surg.-Com..........(X) D. R. F. Bertram, OBE MB, BS, DMR................ 18 Apr 44 (Senior Medical Officer)		Surg. Com..........R. G. Anthony, MB, chB.... 30 Jan 45 (Medical Officer-in-Charge)	
Surg. Lieut.-Com.	(A) M. C. Ross, MB, chB... 24 Apr 44 P. H. K. Gray, MRCS, LRCP 4 Dec 44 (act)............................. 11 Oct 43	Surg. Com., R.N.V.R.	(S) H. G. Ungley, VD, MD, BS, LRCP, FRCS............ 28 Aug 44
Surg. Lieut.-Com., R.N.V.R.	(S) B. C. Murless, MB, FRCS MRCS, LRCP................. 11 Oct 43	Surg. Lieut.-Com.	(E) G. S. Irvine, MB, BS, MRCS, LRCP, DLO......... 28 July 44
Tempy. Surg. Lieut., R.N.V.R.	(M) W. E. Clarke, MB, BS... 12 Aug 43 (E) F. G. Hollands, MB, BS, MRCS, LRCP...... 21 July 44 D. H. Teasdale, MB, BS... — Oct 44	Surg. Lieut.-Com., R.N.V.R.	(M) D. H. Anderson, MB, BCh, BAO (act) .... 1 Aug 44
		Tempy. Surg. Lieut., R.N.V.R.	(O) A. C. Higgitt, MB, BS, MRCS, LRCP................. 5 Sept 44 (A) H. A. B. Nicholls, BCh, MRCS, LRCP .... 5 Dec 44
Tempy. Surg. Lieut. (D), R.N.V.R.	V. L. Simons, LDS........... 28 Feb 45	Tempy. Surg. Lieut. (D), R.N.V.R.	M. A. Kettle, LDS........... 12 Apr 44
Tempy. Chaplain, R.N.V.R.	Rev. E. T. Bartlett.......... 13 Oct 43	Tempy. Chaplain R.N.V.R.	Rev. T. L. Meredyth........ 23 June 43
Tempy. Lieut. (S), R.N.R.	C. D. Scott.................. 17 Feb 42	Tempy. Lieut. (S), R.N.R.	H. W. Cooper................ 14 Aug 41
Reserve Nursing Sister	Miss J. M. Barrow........... — Apr 45 Miss D. E. Evans........ 5 May 45 Miss P. M. Gillum........... — Apr 45 Miss P. E. Hayter........... 5 May 45	Reserve Senior Sister	Miss H. M. Paterson (act) — Apr 44
		Reserve Nursing Sister	Mrs. M. S. J. Howes......... 5 May 45 Miss M. C. Lawrence........ 4 May 45 Miss C. G. N. Percy......... — Apr 45 Miss M. Thorburn............ — Apr 45
Tempy. Wt. Wardmaster	W. T. Franklin.............. 4 Oct 43	Tempy. Wt. Wardmaster	H. F. Weller................. 15 Apr 43

The QARNNS were amalgamated with the rest of the Royal Navy in March 2000 using the same officers' ranks although they retain their QARNNS suffix. The photo below shows Lt Cdr Jan Ouvry QARNNS beside me on board the cruiser HMS *Belfast*, the Second World War and Korean War veteran museum ship, moored on the River Thames in London, in November 2009.

Photo Foreword-4

Lt Cdr Jan Ouvry QARNNS and myself whimsically donning her officer's hat.

Jan, since retired, is the granddaughter of Cdr John Ouvry DSO RN, who was first to render safe a German magnetic ground mine on the mudflats at Shoeburyness in November 1939. Ouvry's mine is displayed on board HMS *Belfast* which was severely damaged by a similar mine, also in November 1939, which put her out of action for the best part of three years. The photo was taken after an event commemorating the 70th anniversary of Ouvry's feat which allowed the boffins (slang term for scientists, engineers, or other personnel engaged in technical or scientific research and development) at HMS Vernon (a shore command in Portsmouth, England) to examine the mine and develop ship self-protective measures.

Photo Foreword-5

Museum ship HMS *Belfast* anchored in the River Thames, with the London Tower bridge in the background.
Courtesy of Rob Hoole

Poignantly, the last of the eight British Army hospital ships sunk during the Second World War was HMHS *Amsterdam*. She actuated a German mine off Juno Beach on 7 August 1944. Fifty-five wounded patients were lost as were ten medical staff and thirty crewmembers. Also lost were eleven German prisoners of war. Total losses were 106 souls.

Photo Foreword-6

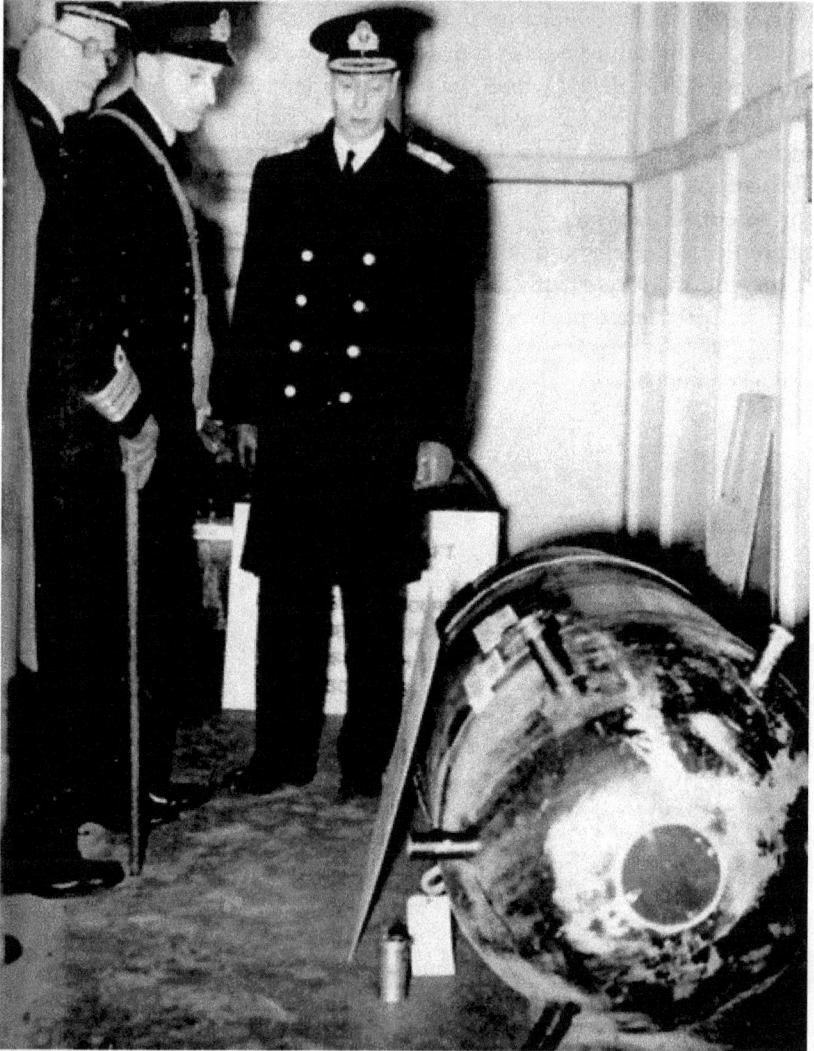

19 Dec 1939 – Lt. Comdr. John Ouvry at HMS Vernon showing King George VI the German magnetic ground mine he rendered safe on the mudflats at Shoeburyness, at the mouth of the Thames Estuary, on 23 November 1939. The projections were intended to prevent the cylindrical mine rolling across the seabed.
Collection of Rob Hoole

Rob Hoole

Foreword

It is almost axiomatic that countries with troops fighting overseas will need to acquire hospital as well as troop ships. Canada pioneered employment of hospital ships in World War I. In that war five hospital ships made 42 voyages to return 28,000 injured servicemen to Canada, victims of her huge commitment to the bloodbath in Europe. These ships included HMHS *Landovery Castle* which was torpedoed and sunk in a well-published atrocity by German submarine *U-86* off southern Ireland on 27 June 1918.

When my friend and this book's author David Bruhn asked me to research and provide perspective regarding the service of Canada's two hospital ships in World War II, I happily took up this task. During my research, I came to realise what a fascinating and important subject hospital ship service is, and amazingly found a link, although somewhat tenuous, to a family story involving one of the ships.

As is described in the text, Canada had two hospital ships in WWII; one the former passenger liner RMS *Letitia*. She and sister ship RMS *Athenia* ran a North Atlantic service between Scotland and eastern Canada for the Donaldson Atlantic Line. On 3 September 1939, *U-30* (Kptlt. Fritz-Julius Lemp) torpedoed the *Athenia* without warning about 250 miles west of Inishtrahull, the most northerly island of Ireland. Sunk with the loss of 117 lives, she was the first ship to fall victim to the Germans in the war.

Photo Foreword-7

SS *Athenia* in Montreal Harbour in 1933.
National Archives of Canada photograph PA-056818

Special friends of my family, Sir Richard Lake and his wife Dorothy were returning from a visit to Britain to their home in Victoria, British Columbia, aboard *Athenia* when she was torpedoed. As a consequence, they spent hours in a life boat before being rescued. My great Aunt Mabel and Godmother, an adopted and contributing member of the Lake household, was looking after Lake children and the home at the time. Only days before the Lakes and their son Dickon, who was studying in England, had visited some of her siblings (also my relatives) in Scotland.

Photo Foreword-8

Sir Richard Lake extreme left, my dad fourth from right, Aunt Mabel third from right, remainder all Lake children, which Aunt Mabel was highly instrumental in raising, except for the child second from left who was a friend.
George Duddy collection

It was a surprise when researching hospital ships, to find this connection and to realise that Captain James Cook of the *Letitia* was the same one who was in command of *Athenia*. Lake was an immigrant farmer and government representative who served initially for the Grenfell district of the then Northwest Territories. He later became the third Lieutenant Governor of the Province of Saskatchewan, serving from 1915 - 1921 before retiring to Victoria British Columbia. I believe that his knighthood, which was still permitted at the time in Canada, was for work he did with the Red Cross Organisation.

Photo Forword-9

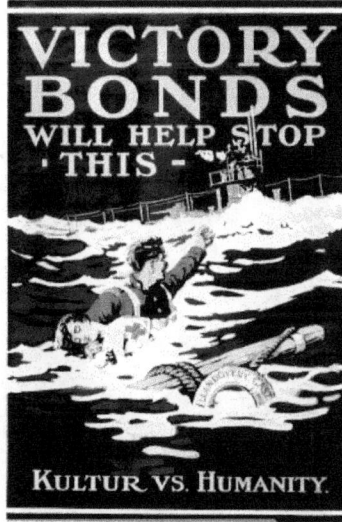

HMHS *Letitia*'s military commander, Lt. Col. A. L. Cornish (left) and her master, Capt. James Cook (right). Cook has previously been the master of *Athenia* when she was sunk, and of *Letitia* in her previous service as a troop transport. At right: 1918 Canadian propaganda poster used the sinking of HMHS *Llandovery Castle* as a focal point for selling Victory Bonds.
Photo credit: http://www.forposterityssake.ca/

As a final personal observation by the writer, it is noted that while it is reported in this book that there were attacks on hospital ships by German aircraft, there were not reported attacks on them by German submarines. I like to think this was because of the widespread publicity of the *Landovery Castle* atrocity of WWI. Perhaps the 234 crew and medical staff, including 14 Nursing Sisters, that were lost in WWI paid the price of the safety of crews and patients of hospital ships from submarines in WWII.

George H.S. Duddy, P. Eng Ret.
White Rock, British Columbia, Canada

Acknowledgements

Sanctuary Not Certain is my thirty-sixth and final book. Accordingly, I would like to recognize, and have identified herein, nearly two score individuals who were kind enough to pen forewords, and/or assist me in other ways from the publishing of *Ready to Answer All Bells* in 1997 to this book, *Sanctuary Not Certain*, in 2025.

Ready to Answer All Bells is devoted to shipboard engineering; and *Sanctuary Not Certain* to Allied Army hospital ships (as well as a few Navy ones) in the European/Middle East Theatres in World War II. Twenty-eight of the intervening books are of the naval history genre. Four others—*Toe the Mark*, *Stride Out*, *Distant Finish*, and *Beavers*—take up competitive running. The remaining two books, *Land Yacht Seaward* and *Stand Easy*, are intended for readers interested in building a mahogany plywood camper (complete with portholes) for a small, lightweight truck; and/or creating a British pub in their garage.

To avoid bogging down readers, I will only highlight people who were most involved in many books over several years. The first is Craig Scott, the owner and CEO of Heritage Books, who published all of my books, except for the initial one. The emphasis of my books has been on recognizing the important contributions of largely unsung ships and their crews, not necessarily material most profitable to a publisher. But for his similar interest, much information made available to readers would likely have otherwise been consigned to the dustbin of history.

Photo Acknowledgements-1

Craig Roberts Scott
Courtesy of Craig Scott

Heritage Books

Founded in 1976 by Laird and Marlene Towle

Purchased In 2002 by Craig R. Scott

The brilliant and very prolific maritime and aviation artist Richard DeRosset also shared my passion and created original paintings for the covers of all of my naval history books, and for this one. Debbie Riley, Heritage Books' editor and book designer, produced the covers using his fine art. Acquisition editor Leslie Wolfinger shepherded the books from acquisition through the publishing process.

Photo Acknowledgements-2

Left: Richard being honored at the San Diego Veterans Museum and Memorial Center. Right: Richard at work in his studio, while a young admirer looks on.
Courtesy of Richard DeRosset

Photo Acknowledgements-3

Rob Hoole on board a *Ton*-class minehunter off Gibraltar in 1984 and, more recently, with Cdr. Martin Mackey RN, Superintendent of Defence Diving, in front of the Vernon Mine Warfare & Division Monument at Gunwharf Quays (formerly HMS Vernon) in Portsmouth.
Courtesy of Rob Hoole

Following the publishing of *Wooden Ships and Iron Men, Vol. I*, in 2006, I contacted retired Royal Navy mine clearance diving officer Rob Hoole. Rob is the founding Vice Chairman and Webmaster of the Royal Naval Mine Minewarfare & Clearance Diving Officers' Association, and has served in this role since the organisation's inception. I told him about the book, and asked him to make members aware of it. This initial contact launched a long friendship. Since then, Rob has reviewed and provided much RN-related material for book manuscripts, penned forewords, and co-authored six books with me.

Photo Acknowledgements-4

Left: George Duddy in front of an old U.S. Army Miki type tug, 2015. Right: George in Glacier Bay aboard the MS *Volendam* during an Alaskan cruise in 2019. A contributor and U.S. Military & Naval Vessel Correspondent for Nauticapedia.ca Project, he is sporting an organisation ballcap.
Courtesy of George Duddy

Nearly ten years ago, Canadian George Duddy contacted me after having read my book *MacArthur and Halsey's "Pacific Island Hoppers."* George is a retired professional engineer with a keen interest in the maritime history of Western Canada and the Arctic, which includes that of the commercial and private service of former U.S. Navy ships. He was seeking information about the hulk of a former USN wooden-hulled, small coastal transport (APc) resting on the bottom in a river slough near his home. This liaison also began a long-lasting friendship and, since then, George has voluntarily served as content editor for at least two dozen of my books, and has co-authored one. His own book *Called by the North*, the subtitle of which is *Extraordinary Adventures of the Fur Trade, Shipbuilders, Navigators and Traders in Northwestern Canada and Alaska*, describes its gripping subject matter, was published in 2022.

Rob also greatly assisted me by providing introductions to flag rank and other senior active or retired Royal Navy and Royal Australian Navy officers. These individuals lent to readers via their forewords, high-level perspective and richness, complementing book subject material, stemming from their deep knowledge of their own and other Commonwealth navies.

Two of these individuals, Commodore Hector Donohue AM RAN (Rtd.) and Rear Adm. Allan du Toit AM RAN (Rtd.), penned forewords, generously allowed my use of material from their own books and articles, and also researched/wrote original material for use in my books.

Photo Acknowledgements-5

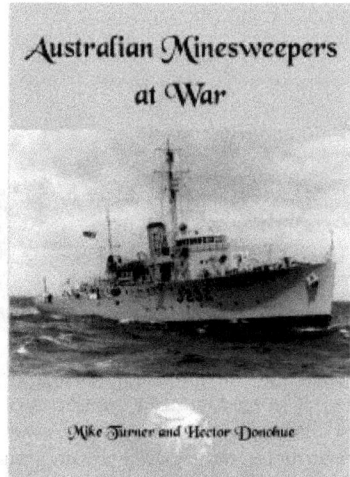

Left: Commodore Hector Donohue, AM RAN (Rtd.). Right: cover of *Australian Minesweepers at War*, one of the several books that he has authored or co-authored. Courtesy of Commodore Donohue

Commodore Donohue began his career in the RAN in 1955 as a seaman officer and subsequently sub-specialised as a clearance diver and torpedo and anti-submarine officer. His service in the RAN included command of the destroyer escort HMAS *Yarra* and the guided missile frigate HMAS *Darwin*. Ashore, he held a number of senior positions in Defence policy and force development prior to retirement in mid-1991.

He has authored and co-authored many books and articles, including the books: *From Empire Defence to the Long Haul Post-war Defence Policy and Its Impact on Naval Force Structure Planning 1945-1955*; *United and Undaunted - the First 100 Years*; *Mines, Mining and Mine Countermeasures*; and *Australian Minesweepers at War*.

Commodore Hector Donohue and Rear Adm. Allan du Toit AM RAN (Rtd.) share being retired Royal Australian Navy admirals (commodore is a 1-star admiral), but du Toit began his lengthy, and very distinguished naval career as a member of the South African Navy. Accordingly, he also has much personal experience and extensive knowledge of it, and provided related material and photographs from his collection for use in several of my books where that nation's involvement was covered.

Photo Acknowledgements-6

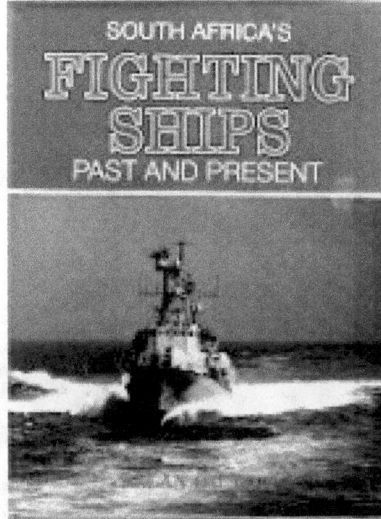

Left: Rear Adm. Allan du Toit, AM RAN (Retired). Right: cover of one of Admiral du Toit's books.
Courtesy of Rear Admiral du Toit

Rear Admiral Du Toit retired from naval service in 2016 after forty years combined service in two Commonwealth navies. Born and reared in South Africa, he entered the South African Navy in 1975 and joined the Royal Australian Navy in early 1987. As a Warfare Officer at sea, du Toit served in mine counter measures vessels, submarines, frigates, destroyers and amphibious ships, commanding at each rank from Lieutenant to Rear Admiral. Flag officer assignments included two Middle East coalition command appointments. His final appointment was as Australia's Military Representative to NATO in Brussels.

Du Toit authored two books, *Ships of the South African Navy* and *South Africa's Fighting Ships - Past and Present* during his operationally-focused career, and over the years he has also contributed to various other books, journals and naval history conferences. Keenly interested in academics, he earned his doctorate degree from the University of New South Wales (UNSW) Canberra, the focus of his research being the defence of the Cape Sea Route during the Cold War.

Once research, writing, and inclusion of material provided by subject matter experts are completed, editors become extraordinarily important to the products of authors and writers. I have been privledged to have George Duddy serve as content editor for a number of reasons. Being an engineer (and technical writer near the end of his career), a maritime expert, citizen of a Commonwealth country, and member of the generation that precedes mine, he lent unique attributes to his review/rework of draft manuscripts.

Photo Acknowledgements-7

Lynn Marie Tosello
Courtesy of Lynn Marie Tosello

I was very fortunate to become acquainted with Lynn Marie Tosello a number of years ago and, since then, she has edited over two dozen of my books. I particularly value her work. In addition to having the worldliness, critical eye, and knowledge of syntax and prose, one would expect of a topline editor, it benefited me that she was not an expert on the Navy and naval matters, and could identify to me where sufficient explanation about some maritime topic was lacking.

THE SCORES OF DISTINGUISHED, ACCOMPLISHED INDIVIDUALS WHO LENT THEMSELVES TO ONE OR MORE BOOKS

Coauthors

Al Baeta – *Beavers*
George Duddy – *Rarely Idle*
Stephen Ekholm – *Salvation from the Sky*
Rob Hoole – *Enemy Waters*; *Home Waters*; *Nightraiders*; *Rarely Idle*;
  Send Some King's Ships; *Stream Gear*
Jack Leydig – *Distant Finish*
Richard S. Mathews – *On the Gunline*

Forewords

Ready to Answer All Bells
Adm. James R. Hogg, USN (Ret.)
Rear Adm. James F. Amerault, USN

Sanctuary Not Certain
Commodore Hector Donohue, AM RAN (Rtd.)
Lt. Cdr. Rob Hoole, RN (Rtd.)
George H. S. Duddy, P. Eng. (Ret.)

Naval History Books
The Hon. Charles L. Cragin
Rear Adm. Allan du Toit AM RAN (Rtd.)
Rear Adm. Stephen K. Chadwick, USN (Ret.)
Rear Adm. Charles F. Horne III, USN (Ret.)
Rear Adm. Paddy A. McAlphine CBE RN (Rtd.)
Rear Adm. Paul J. Ryan, USN (Ret.)
Rear Adm. Christopher Weaver, USN (Ret.)
Commodore Hector Donohue, AM RAN (Rtd.)
Capt. Gary T. Carter, USN (Ret.)
Capt. Christopher O'Flaherty, RN
Capt. Steven C. Saulnier, USN (Ret.)
Capt. Richard Tarbuck, USN (Ret.)
Capt. John D. White II, USN (Ret.)
Cdr. Lee M. Foley, USN (Ret.)
Cdr. Fraser M. McKee, RCNR (Rtd.)
Cdr. William I. Milwee Jr., USN (Ret.)
Cdr. Ron Swart, LDO, USN (Ret.)
Cdr. Larry Wahl, USN (Ret.)
Lt. Cdr. Rob Hoole, RN (Rtd.)
Dr. Edward J. Marolda

Dr. Salvatore R. Mercogliano
Iain G. Cameron
George H.S. Duddy, P. Eng. (Ret.)
August Felando
Kemper Goffigan III
Roland Nino Martinez
Dwight R. Messimer
Melville Owen
David Rajkovich
Gordon Smith

Toe the Mark / Stride Out / Distant Finish / Beavers
Dr. Britton Brewer
Merill (Cray) Creagh
Bob Darling
Bob Deis
Kim Ellison
Dr. Laura de Ghetaldi
Bill Gregg
Jack Leydig
Joe Mangan
Dr. Walt Schafer
Jean Snuggs

Land Yacht Seaward / Stand Easy
Iain G. Cameron
George H.S. Duddy, P. Eng. (Ret.)
Lt. Cdr. Rob Hoole, RN (Rtd.)
Lynn J. Salmon

Editing / Book Design / Typesetting
Carolyn L. Barkley
Lt. Cdr. Pelham Boyer, USN (Ret.)
Dr. Robert Meindl
Jo-Ann Parks
Debbie Riley
Lynn Marie Tosello

Preface

Sure we rough it. But compared to the way you men are taking it we can't complain, nor do we feel that bouquets are due us ... it is to you we doff our helmets. To every G.I. wearing the American uniform — for you we have the greatest admiration and respect.

—Lt. Frances Y. Slanger was one of the Army nurses who signed a letter written to *Stars and Stripes*. Seventeen days later, on 21 October 1944, she died of wounds caused by the shelling of her tented hospital area in Belgium.[1]

To all Army nurses overseas: We men were not given the choice of working in the battlefield or the home front. We cannot take any credit for being here. We are here because we have to be. You are here because you felt you were needed. So, when an injured man opens his eyes to see one of you ... concerned with his welfare, he can't but be overcome by the very thought that you are doing it because you want to ... you endure whatever hardships you must to be where you can do us the most good.

—Through the same newspaper, hundreds of soldiers replied; their heartfelt words included those above.[2]

Photo Preface-1

U.S. Army Nurse Corps Pin

Women Officers of the Armed Forces on the Lincoln Memorial, Washington, D.C., circa 1942-43. Left to right: U.S. Army nurse, U.S. Navy nurse, U.S. Navy WAVE (Women Accepted for Volunteer Emergency Service), and U.S. Army WAAC (Women's Army Auxiliary Corps).
National Archives photograph #80-G-K-13431

The title of this book, *Sanctuary Not Certain*, alludes to the fact that during World War II, German/Italian aircraft purposefully bombed and sank thirteen Allied hospital ships in the European Theatre. A fourteenth hospital ship was sunk by a German mine in the English Channel off Juno beach, during the Normandy invasion. The well-known mariners' saying "any port in a storm," is a reference to searching for/obtaining sanctuary from intolerable/dangerous conditions at sea. The loss of the following ships to the causes cited evidences that wounded, sometimes gravely so, servicemembers could not be assured of safety aboard hospital ships, and neither could the vessels' crews nor the doctors, nurses and other medical staff aboard to provide patient care.

Norwegian
Hospital Ship *Dronning Maud* (1 May 1940 – German aircraft)

Greek
HS *Attiki* (11 April 1941 – German aircraft)
HS *Esperos* (21 April 1941 – German aircraft)
HS *Sokratis* (22 April 1941 – German aircraft)
HS *Andros* (23 April 1941 – German aircraft)
HS *Ellenis* (21 April 1941 – damaged by German aircraft, grounded by master, sunk later that month)

British
Hospital Carrier HMHS *Maid of Kent* (21 May 1940 – German aircraft)
Hospital Carrier HMHS *Brighton* (24 May 1940 – German aircraft)
Hospital Carrier HMHS *Paris* (2 June 1940 – German aircraft)
Hospital Ship HMHS *Ramb IV* (10 May 1942 – German aircraft)
Hospital Ship HMHS *Talamba* (10 July 1943 – Italian aircraft)
Hospital Ship HMHS *Newfoundland* (13 September 1943 – German aircraft)
Hospital Carrier HMHS *St. David* (24 January 1944 – German aircraft)
Hospital Carrier HMHS *Amsterdam* (7 August 1944 – German mine)

BRITISH HOSPITAL SHIPS / HOSPITAL CARRIERS
Hospital ships served the British Army in evacuating the wounded back to Britain. Their numbers were supplemented by hospital carriers which had shallower draughts and could go closer inshore to facilitate swifter evacuation of casualties. (As a general rule, hospital ships were converted passenger liners or troop ships, and hospital carriers smaller converted cross channel ferries.) Hospital ships/ carriers were assisted by water ambulances (boats) with flat bottoms that could carry stretchers and land ashore.

HMHS (His Majesty's Hospital Ship) prefaced the names of both hospital ships and hospital carriers, and articles about hospital carriers often identified them as hospital ships. During World War II, the British had 42 hospital ships/hospital carriers; other Commonwealth countries, collectively, 7 hospital ships (identified by shading in the table). A lengthier table in Appendix A, identifies to readers whether the below named vessels were hospital ships or hospital carriers.

WWII Commonwealth Hospital Ships/Hospital Carriers (49)

HMHS *Aba*	HMHS *Isle of Jersey*	HMAHS *Oranje*
HMHS *Amarapoora*	HMHS *Isle of Thanet*	HMHS *Oxfordshire*
HMHS *Amsterdam*	HMHS *Karapara*	HMHS *Paris*
HMHS *Atlantis*	HMHS *Karoa*	HMHS *Prague*
HMHS *Brighton*	HMHS *Lady Connaught*	HMHS *Ramb IV*
HMHS *Cap St Jacques*	HMCHS *Lady Nelson*	HMHS *St Andrew*
HMAHS *Centaur*	HMHS *Leinster*	HMHS *St David*
HMHS *Chantilly*	HMCHS *Letitia*	HMHS *St Julien*
HMHS *Dinard*	HMHS *Llandovery Castle*	HMHS *Tairea*
HMHS *Dorsetshire*	HMHS *Maid of Kent*	HMHS *Talamba*
HMHS *Duke of Argyll*	RFA *Maine*	HMHS *Takliwa*
HMHS *Duke of Lancaster*	HMAHS *Manunda*	HMHS *Tjitjalengaka*
HMHS *Duke of Rothesay*	HMNZHS *Maunganui*	HMHS *Vasna*
HMHS *El Nil*	HMHS *Naushon*	HMHS *Vita*
HMHS *Empire Clyde*	HMHS *Newfoundland*	HMAHS *Wanganella*
HMHS *Gerusalemme*	HMHS *Ophir*	HMHS *Worthing*
HMHS *Isle of Guernsey*		

HMHS: His Majesty's Hospital Ship
HMAHS: His Majesty's Australian Hospital Ship
HMCHS: His Majesty's Canadian Hospital Ship
HMNZHS: His Majesty's New Zealand Hospital Ship

BRITISH MEDICAL / NURSING PERSONNEL

Here borrow my comb, and try not to look like survivors!

—Guidance from a matron to her junior nursing sisters following
several hours in Mediterranean water, clinging to wreckage after
their hospital ship was sunk, and having been brought
ashore in Oran where their bedraggled uniforms
were exchanged for men's clothing.[1]

Photo Preface-2

Hospital ships were distinctive, usually painted white with horizontal green stripes on their hulls and displaying large red crosses but that did not make them immune from attack, as nurses of the QAs (Queen Alexandra's Imperial Military Nursing Service, QAIMNS) learned.

These brave women sailed back and forth across the English Channel many times during the evacuation of troops from Dunkirk in May-June 1940, enduring bombs, mines and attacks by the Luftwaffe.

They rarely stopped to think about the danger they faced, and instead at all times cared for the wounded men entrusted to their care.

Poster enjoining British civilian nurses to join the Queen Alexandra's Imperial Military Nursing Service Reserve (QAIMNSR).

On board British Army hospital ships, female nurses of the Queen Alexandra's Imperial Military Nursing Service (QAIMNS) worked alongside male doctors of the Royal Army Medical Corps (RAMC) providing care for embarked wounded soldiers. When caring for the more complex cases, nurses required assistance from male orderlies, placing fit military men in the charge of women, contrary to the then normal social conventions regarding gender in British society.[2]

As was the case in WWI, the engagement of female nurses in front-line duty in the war created opportunities for British registered nurses that their peacetime work could not, and that were not present in the hospitals on the home front. Nursing was seen as the epitome of female work, yet from 1941 Nursing Sisters were commissioned officers in the British Army. Moreover, in war zones where they were often the only women allowed, their traditional nurses' uniform was replaced with male battle-dress.[3]

The challenges which nurses faced were profound. At the outbreak of war, there was a peacetime allocation of 624 members of the QAIMNS serving in Military Hospitals at home and abroad. This number was greatly increased with the mobilisation of QAIMNS(R) and TANS (Territorial Army Nursing Service) members.[4]

In 1938 when war in Europe was anticipated, Matrons in civilian hospitals, particularly teaching hospitals, recommended suitable trained nurses for entry into the QAIMNS(R). Such nurses were interviewed at the War Office and given a sealed packet which bore the words "Open only in the event of war" (which presumably contained information regarding their roles in the likely event of mobilisation).[5]

A vast majority of the QAIMNS(R) Nursing Sisters who went to war between 1939 and 1945 had no previous military experience. Additionally, few had ever been abroad, let alone worked overseas; and many had lived the sheltered lives of young respectable women before moving from the parental home to the hospital Nurses' Home, to field hospitals overseas, or aboard hospital ships.[6]

NURSES OF OTHER COMMONWEALTH COUNTRIES

Photo Preface-3

Middle East: British Mandate of Palestine, Palestine, Gaza. Matron Sage of the 2nd Australian General Hospital at Gaza Ridge has accumulated quite a bevy of beauty in her nursing staff – in fact it's something of an inducement to report sick these days. Australian War Memorial photograph 004013 with quoted caption

Australia, New Zealand, and Canada Nursing Sisters likewise provided care for wounded soldiers in field hospitals, aboard their country's few, but very important hospital ships, and in hospitals at home.

SEQUENCE OF CARE, FIELD TO HOME HOSPITALS

It is important to note here that hospital ships, the subject of this book, were but one part (albeit an important one) of organised military "chains of evacuation" used to treat and return wounded to their units or to take them home for additional recuperation or to discharge them from the military if their injuries precluded continued service.

U.S. Army Medical Department's practices, first utilised and honed in North Africa, are detailed in Chapter 6. The following bulletised summary information introduces this patient chain of evacuation:

- Mobile field and evacuation hospitals, which closely followed the combat troops, were usually set up in tents and were subject to move at short notice
- Nurses packed and unpacked hospitals each time they moved
- Patients were brought to field hospitals by litter bearers and ambulance drivers
- Doctors and nurses performed triage on arriving patients at the receiving tent to ascertain the severity of their condition
- Those judged strong enough to travel were taken by ambulance to evacuation hospitals farther away from the front lines and near transportation facilities
- Those who needed immediate care went directly into surgery
- Others who needed surgery but were too weak for an immediate operation and could not travel, were sent to the shock ward
- Evacuation hospital doctors operated as necessary on patients sent from field hospitals
- Where facilities were available, critically wounded patients needing specialised treatment were evacuated by air to station and general hospitals
- Other patients were evacuated from the field to station and general hospitals via hospital trains, hospital ships, and aircraft attended by nurses and other medical personnel
- Station hospitals received battle casualties from evacuation hospitals and performed surgery and specialised treatments
- General hospitals were the last step in the evacuation line; patients still requiring diagnosis, specialized lab tests, or long periods of recuperation and therapy were sent to general hospitals
- Upon release, patients were either returned to duty or sent back to the United States

OPERATIONS IN THE EASTERN MEDITERRANEAN

The bulk of *Sanctuary Not Certain* is devoted to combat in the western Mediterranean from the Allied Invasion of North Africa in November 1942 through the Allied Invasion of Southern France in August 1944. The Normandy Invasion in June 1944, is also taken up, the site of the loss of the last Allied hospital ship in the European Theatre.

Map Preface-1

Eastern Mediterranean

Coverage of warfare in the eastern Mediterranean is limited to the role of Australian hospital ships, described by Commodore Donohue in his foreword to the book, and the loss of the five Greek hospital ships in April 1941 to vicious attacks by the Luftwaffe (German Air Force).

Photo Preface-4

Sick and wounded Australian and New Zealand troops from the Middle East were brought home safely on the Australian hospital ship *Oranje*. In appreciation of the good work done by Dutch personnel and Australian nurses and staff, on arrival in New Zealand, the New Zealand Government gave them a three-day trip to the Chateau Tongariro in the snow country. Transport, accommodation, meals, snow and sporting equipment were provided at government expense. After the long and hazardous sea voyage this was a little bit of heaven. After some strenuous outdoor exercise these Nursing Sisters quench their thirst with natural ice blocks.
Australian War Memorial photograph 150875

OPERATIONS IN THE WESTERN MEDITERRANEAN

Map Preface-2

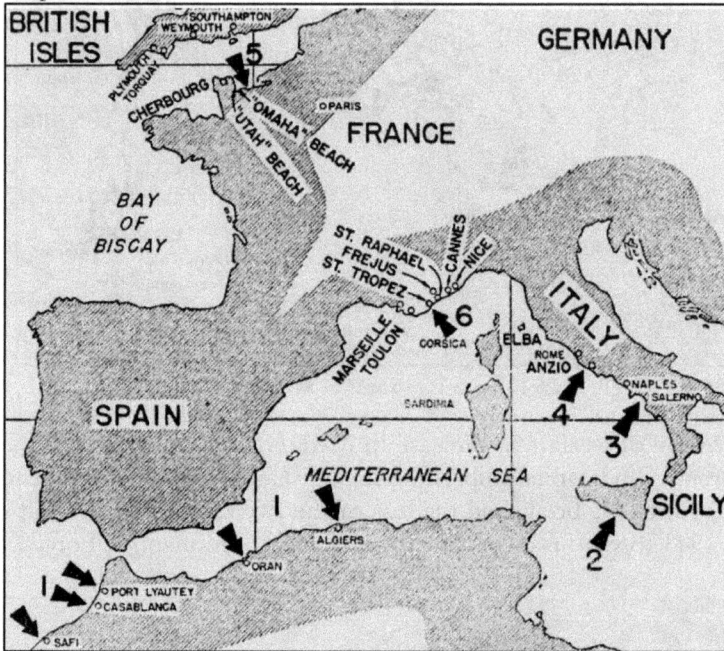

Naval Operations in North Africa, France, and the western Mediterranean
https://www.ibiblio.org/hyperwar/USN/USNatWar/img/USN-King-p83.jpg

1. North African Landings 8 Nov 42	4. Anzio Landings 22 Jan 44
2. Sicilian Landings 10 Jul 43	5. Normandy Invasion 6 Jun 44
3. Italian Landings 9 Sep 43	6. Invasion of Southern France 15 Aug 44

The preceding map nicely illustrates the series of major operations in which U.S. Army hospital ships and those of Allied navies, most notably the Royal Navy, took part in the western Mediterranean between November 1942 and August 1944. As shown, they began with the invasion of North Africa; then proceeded counter-clockwise inside the sea basin with invasions of Sicily, Salerno, and Anzio; and concluded in southern France. Wounded from the Allied Normandy Invasion in northwestern France, were evacuated by British hospital carriers across the English Channel to Britain.

Since no American hospital ships were available for evacuation operations in the European Theatre, four British hospital carriers were dedicated to United States support during the D-Day landings. They were HMHS *Dinard*, *Lady Connaught*, *Naushon*, and *Prague*. As an example of their employment, on the night of 8-9 June, the *Lady Connaught* landed by boats at Utah Beach the U.S. Army First Medical Detachment, consisting of the following personnel:

- Station and litter bearer platoons of the 502nd and the 427th Medical Collecting Companies (31st Medical Group)
- Six Surgical Teams of the 4th Auxiliary Surgical Group
- One Advance Depot Platoon (31st Medical Depot Company)
- Six Medical Corps Officers of the 662nd Medical Clearing Company (134th Medical Group)
- Ten Liaison Officers from various medical units, including one Officer of the 9th Troop Carrier Command[7]

Despite *Lady Connaught*'s rated capacity of approximately 300 casualties, she took aboard 400 wounded personnel during the day, then sailed for the United Kingdom that same evening.[8]

U.S. ARMY HOSPITAL SHIPS

During World War II the Army Transport Service operated a total of 24 hospital ships. The ships were operated by civilian Merchant Marine employees of the Transportation Corps. The Hospital Unit embarked aboard each ship (staffed by Army Medical Department doctors, nurses, and enlisted male orderlies) functioned as an afloat station hospital.[9]

Most of the twenty-four hospital ships were former passenger liners/troopships, but among them were six newly-built Liberty ships converted for hospital use. These ships were the USAHS *Blanche F. Sigman*, *Dogwood*, *Jarrett M. Huddleston*, *John J. Meany*, *St. Olaf*, and *Wisteria*.

The hospital ships highlighted by shading in the following table earned Europe-Africa-Middle East Campaign Medals during the war.

WWII U.S. Army Hospital Ships (24)

USAHS	Built	Speed/ Patient Capacity	First Voyage / Departure Date	Embarked Hospital Unit
Acadia	1932	18 kts/ 787	New York to N. Africa – 5 Jun 43	204th Medical Hospital Ship Co.
Seminole	1907	14 kts/ 454	New York to N. Africa – 20 Sep 43	
Shamrock	1906	14 kts/ 543	New York to N. Africa – 4 Sep 43	202nd Medical Hospital Ship Co.
Algonquin	1926	15 kts/ 454	New Orleans to N. Africa – 2 Feb 44	203rd Medical Hospital Ship Co.
Thistle	1921	14 kts/ 455	New York to N. Africa – 8 Apr 44	206th Medical Hospital Ship Co.

Chateau Thierry	1920	16 kts/ 484	Boston to N. Africa – 5 Mar 44	208th Medical Hospital Ship Co.
Ernest Hinds	1918	12 kts/ 288	Charleston to Italy – 14 Jul 44	
Dogwood	1943	11 kts/ 592	New York to UK – 21 Jul 44	218th Medical Hospital Ship Co.
Larkspur	1901	10 kts/ 592	Charleston to UK – 31 Aug 44	209th Medical Hospital Ship Co.
St. Mihiel	1920	16 kts/ 504	Boston to N. Africa – 10 May 44	
Wisteria	1943	11 kts/ 588	New York to UK – 16 Jul 44	219th Medical Hospital Ship Co.
John L. Clem	1918	12 kts/ 286	Charleston to N. Africa – 15 Jun 44	
Marigold	1920	12 kts/ 758	Charleston to Italy – 19 Jul 44	212th Medical Hospital Ship Co.
St. Olaf	1942	11 kts/ 586	New York to UK – 12 Aug 44	217th Medical Hospital Ship Co.
Emily H. M. Weder	1920	13 kts/ 738	New York to Italy – 12 Jul 44	211th Medical Hospital Ship Co.
Jarrett M. Huddleston	1942	11 kts/ 582	New York to UK – 2 Sep 44	
John J. Meany	1942	11 kts/ 582	New York to Italy – 27 Jul 44	
Blanche F. Sigman	1943	11 kts/ 590	New York to UK – 7 Jul 44	220th Medical Hospital Ship Co.
Charles A. Stafford	1918	16 kts/ 706	New York to UK – 21 Sep 44	
Louis A. Milne	1919	12 kts/ 952	Boston to UK – 19 Mar 45	200th Medical Hospital Ship Co.
Ernestine Koranda	1919	12 kts/ 722	New York to UK – 13 Apr 45	
Aleda E. Lutz	1931	16 kts/ 778	New York to UK – 18 Apr 45	
Frances Y. Slanger	1927	19 kts/ 1,628	New York to UK – 30 Jun 45	235th Medical Hospital Ship Co.
Republic	1907	12 kts/ 1,242	New Orleans to Southwest Pacific – 4 Sep 45	234th Medical Hospital Ship Co.[10]

Information about these U.S. Army hospital ships, presented in a different format, may be found in Appendix B.

U.S. ARMY NURSES

On 8 September 1939, mere days after Germany's invasion of Poland, President Franklin D. Roosevelt issued a proclamation of a "limited" national emergency because of the war in Europe. There were then 625 regular U.S. Army nurses on active duty. The authorised strength of the

Army Nurse Corps was immediately increased to 949, and ten months later on 30 June 1940, there were 942 regular Army nurses in the Corps. Additionally, there were 15,770 nurses, enrolled in the First Reserve of the American Red Cross Nursing Service, who would presumably be available for service if needed.[11]

Photo Preface-5

Drawings by McClelland Barclay in 1943 of (at left) Lt. Margaret Wheeler, Army Nurse Corps, who volunteered for foreign duty; and at right, of Lt. Willa Lucille Hook, Army Nurse Corps, who joined the U.S. Army in 1939.
Naval History and Heritage Command photographs #85-236-AE and #85-236-AU

On 7 December 1941, Japanese planes bombed Pearl Harbor. Within forty-eight hours, the United States declared war against Japan, Germany, and Italy. There were fewer than seven thousand Army nurses on active duty when the United States entered World War II; but by 30 June 1942, there were 12,475 Army nurses on active duty.[12]

On 8 November 1942, nurses landed in North Africa on the same day that assault troops stormed ashore. They were members of the 48th Surgical Hospital, later reorganised as the 128th Evacuation Hospital.

The first class of Army Nurse Corps flight nurses was graduated by the School of Air Evacuation at Bowman Field, Kentucky, on 18 February 1943. By virtue of being the honor graduate (1st in her class), 2nd Lt. Geraldine Dishroom earned the first flight nurse wings. Since there was, as yet, no official insignia, Brig. Gen. David Grant, Air Surgeon and guest speaker, unpinned his wings and pinned it to her uniform. Dishroom was with the first air evacuation team to land on Omaha Beach after the Normandy invasion on 6 June 1944.[13]

Photo Preface-6

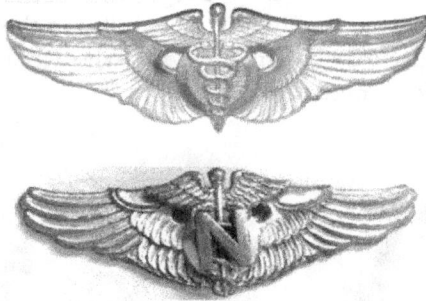

Top: U.S. Army Air Force Flight Surgeon wings, and bottom Flight Nurse wings.

On 19 July 1943, basic training centres, established to provide military orientation for Army nurses before their first duty assignment, were opened at Fort Devens, Massachusetts; Fort Sam Houston, Texas; and Camp McCoy, Wisconsin. At these facilities, nurses were oriented to military nursing and other subjects, such as how to prepare for gas injuries, bivouac in the field, seek foxholes for cover, and purify water.[14]

Seven months later, on 27 January 1944, Army nurses waded ashore on the Anzio beachhead in Italy five days after the troop landings on 22 January. Six Army nurses lost their lives during enemy bombing attacks in early February.[15]

On 8 May 1945 V-E (Victory in Europe) Day was proclaimed following the surrender of Germany the preceding day. When the war in Europe ended there were more than fifty-two thousand Army nurses on active duty serving in 605 hospitals overseas and 454 hospitals in the United States.[16]

Two hundred one Army nurses died during the war, sixteen as a result of enemy action. More than 1,600 nurses were decorated for meritorious service and bravery under fire. Decorations included the Distinguished Service Medal, Silver Star, Legion of Merit, Distinguished Flying Cross, Soldier's Medal, Bronze Star Medal, Purple Heart, Air Medal, and Army Commendation Medal.[17]

LIMITATIONS OF THE BOOK; CLOSE RELATIONSHIPS BETWEEN U.S. / COMMONWEALTH HOSPITAL SHIPS

Information about U.S. hospital ships is scarce, because unlike Navy leadership which required its ships to maintain war diaries, there was no such requirement for Army hospital ships operated by civilian mariners.

Moreover, the attention of the top Army surgeon who commanded the Hospital Unit aboard each ship was largely, rightly devoted to patient care. Fortunately, Lt. Col. Thomas B. Protzman, MC—commanding officer of the 204th Medical Hospital Ship Company aboard USAHS *Acadia*—kept a personal diary which was available for reference. Details he recorded on a daily basis provide a valuable "birds eye view" of medicine aboard a hospital ship.

Returning to the theme of "any port in a storm," it's likely that hurt servicemen had few qualms about which countries or service's medical staff provided them necessary care. As will be discussed in the book, British hospital ships evacuated as necessary American wounded, and U.S. Army hospital ships did likewise for British patients.

As a nod to America's Commonwealth allies in World War II, the British spelling of particular words is used throughout the book. (They also apply to the Dominions generally.) The primary differences are mostly the addition of the letter "u" in some words, and the use of "s" instead of "z" in others.

British	American	British	American
armour	armor	manoeuvre	maneuver
authorise	authorize	metre	meter
defence	defense	organisation	organization
despatch(es)	dispatch(es)	programme	program
draught	draft	publicise	publicize
endeavour	endeavor	theatre	theater
favourable	favorable	vigourous	vigorous
harbour	harbor		

ROYAL NAVY RANKS

The RNVR (classic, wartime reservists known as 'Saturday night sailors') were gentlemen trying to be sailors.

The RNR (professional seamen and part-time Navy officers) were sailors trying to be gentlemen.

and The RN (regular Navy officers) were neither trying to be both.

—Old saying in the Royal Navy, courtesy of Rob Hoole

This introduction closes with an overview provided by Rob Hoole, of the differences between Royal Navy, Royal Naval Reserve, and Royal Naval Volunteer Reserve personnel assigned to Royal Navy ships and to some British hospital ships.

The Naval Reserve Act of 1859 established the Royal Naval Reserve (RNR) as a reserve of professional seamen from the British Merchant Navy and fishing fleets, who could be called upon during times of war to serve in the regular Royal Navy. In 1862, the RNR was extended to include the recruitment and training of reserve officers, who wore on their uniforms a unique and distinctive lace consisting of stripes of interwoven chain. The Royal Naval Volunteer Reserve (RNVR)—the so-called "wavy Navy"—was created in 1903.

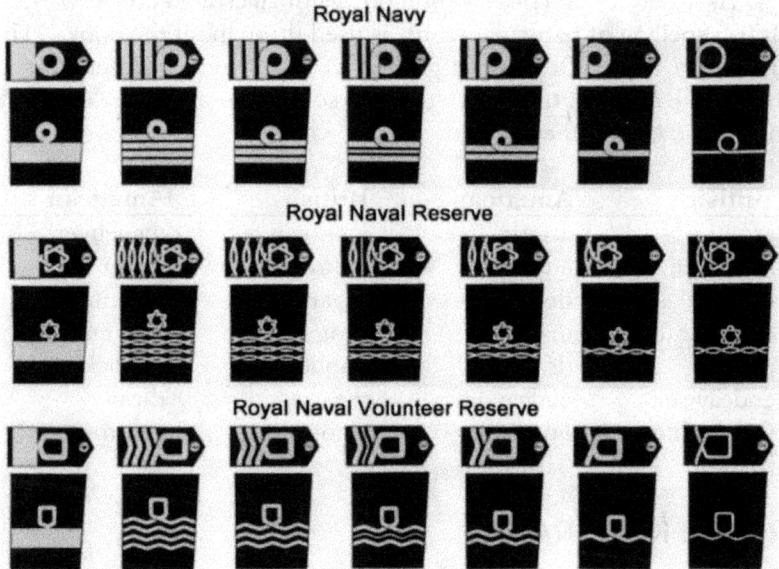

World War II British Naval Officer shoulder boards and sleeve insignia

While the Royal Naval Reserve was composed of personnel from the merchant marine and fishing communities, members of the Royal Naval Volunteer Reserve came from other civilian backgrounds. Another difference was the gold braid officers wore on their sleeves to denote rank. Those of the RNR were in intersecting waved pairs, while the pattern of RNVR braid was single waved lines—thus the RN moniker "wavy Navy" when referring to the latter officers.

Passenger Liners to Hospital Ships

The Navy and Army operated hospital ships with different purposes - Navy hospital ships were fully equipped hospitals designed to receive casualties direct from the battlefield and also supplied logistical support to frontline medical teams ashore, while Army hospital ships were essentially equipped to evacuate patients from forward area Army hospitals to rear area hospitals (or from those to the home port) and were not equipped or staffed to handle large numbers of direct battle casualties.

—Journal of Marine Medical Society 2017.

Photo 1-1

HMAHS *Oranje*, a Netherlands ocean-liner which was loaned to the Australian and New Zealand governments for the duration of WWII. She was a fully equipped hospital ship for Australian and New Zealand forces from June 1941.
Australian War Memorial photograph P01189.002

Before delving into Allied WWII hospital ship operations (which sadly included the sinking of 13 hospital ships in the European Theatre at the hands of ruthless enemy air forces, and 1 to an enemy mine), a little precursory information about these mercy ships is in order. This chapter is devoted to the conversion of Australian and New Zealand passenger liners to hospital ships. However, the considerations and challenges involved were largely applicable to other Allied hospital ships. Two references were used to compile this material:

- AWM, Second World War Official Histories, Series 5 – Medical, Volume IV, Chapter 38, Hospital Ships, 1961
- NZETC, Official History of New Zealand in the Second World War 1939-45, Medical Services in New Zealand and the Pacific, Part VI – Hospital Ships, 1958

FACTORS INVOLVED IN SHIP CONVERSION

In most instances in WWII, it was more expedient and practical to obtain hospital ships by converting existing passenger ships rather than by new construction. Converting a merchant liner to a hospital ship is not a simple task and to understand why the expertise from several agencies is required, the general control of the ship needs to be recognised. A hospital ship in World War II was regarded as a military unit whose senior medical officer was the officer commanding and had overall responsibility for administering those onboard. The Navy carried out maintenance and determined the ship's movements from the point of view of security and safety, and the Merchant Navy was responsible for the sailing of the ship. Consequently, planning for conversion required the combined input and technical expertise from Army medical authorities, Navy and the Merchant Marine. There are several general issues to be considered when selecting a suitable ship for conversion and these are discussed in this chapter. The conversion of the New Zealand hospital ship *Maunganui* will then be described, providing a practical example of what was entailed.

Photo 1-2

Lt. Col. G.R. Halloran, Commander 2/1
Australian Hospital Ship *Manunda*, 30 June 1944.
Australian War Memorial photograph 043457

Photo 1-3

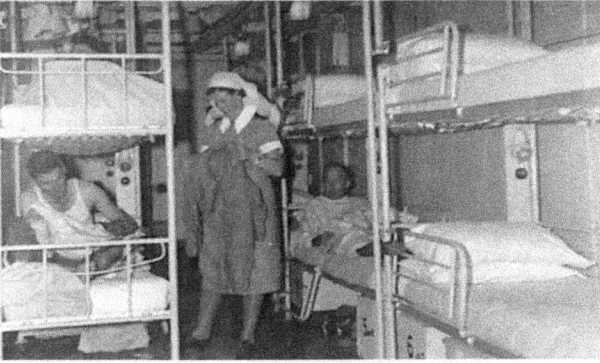

One of the bed sections of a ward on the hospital ship *Wanganella*,
in Port Tewfik, Suez Canal, Egypt, 5 November 1941.
Australian War Memorial photograph 021221

Ship size is important, as the draught of a large ship may be a bar
to full mobility: she may not be able to approach shallow waters or lie
up in shallow anchorage, and this complicates loading and unloading.
Speed should be sufficient to give a quick turn-round, at least 14 knots.
Motor ships have advantages: diesel engine craft have ample electric
power at their disposal and can use exhaust heat (cooling water heated
by the absorption/removal of engine heat) to distill fresh water.

The water supply is important and large quantities must be carried
on board in storage tanks, as it is not practical to distill the vast quantities
required. Average warships need 10 tons a day for each 100 men, but
hospital ships typically use 50 tons daily for each 100 men, with another
50 tons a week for the laundry, a vital factor on a hospital ship. An
emergency boiler unit could produce another 10 to 15 tons a day, but
these smaller distillation plants were not always very successful. The
range of a hospital ship may thus be limited by its water supply, as its
average requirements are not less than 30 to 35 gallons per person a day,
allowing an extra day's ration each week for laundry.

Ventilation is also important. Generally, ships are fitted with
punkah-louvre systems (outlets from ventilation ducting suitable for
high velocity spot cooling applications) but additional ventilation is
needed in some spaces, especially in the wards in the tropics or during
rough weather when air intakes are closed. In the operating theatre some
form of air conditioning, or an exhaust system with special diffusion
arrangements designed to avoid draughts is required. Fully equipped
hospital ships include a special theatre block comprising the theatre, a
plaster room, and an X-ray department, all constructed as one unit.

Photo 1-4

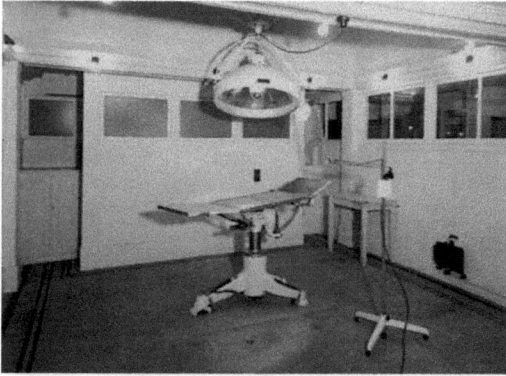

Operating theatre on board the hospital ship *Manunda*.
Australian War Memorial photograph 002396

Adequate lighting is another essential feature, as maximum illumination needs to be secured without glare, preferably using an indirect system. A portable light was found necessary in the theatre, and special care had to be taken in installing it so that it could be operated without being disturbed by ship movement.

The cots followed a standard pattern found satisfactory in other conversions. Both single- and double-tiered types were used. Accommodation in the ships was chiefly in wards. Special wards were provided for orthopedic cases, patients seriously ill, and those with tuberculosis, infectious diseases and mental afflictions. Special observation rooms and strong rooms were provided for psychiatric cases.

Photo 1-5

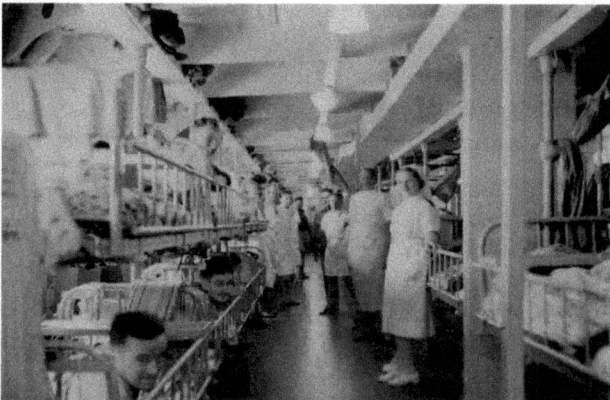

Army Nursing Sisters and orderlies look after their ward and patients.
National Library of New Zealand

Certain other features were common to all hospital ships, such as separate dining accommodation for convalescent patients, and staff, standard diets of four simple types, and sufficient room for occupational therapy and exercise.

Most important were the arrangements for embarkation and disembarkation. Winches, and special stretchers such as the Neil Robertson stretcher and the Stokes litter permitted the handling of the helpless without disturbance, but after their arrival on board, there still remained the problem of their transfer to wards.

Photo 1-6

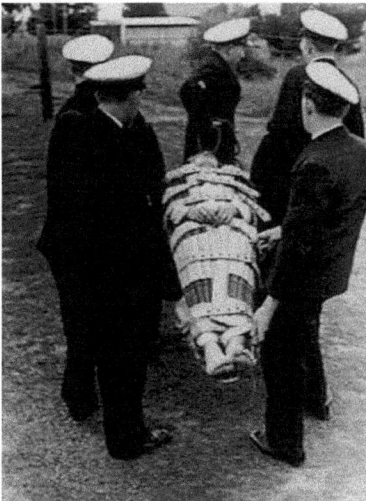

At left: Royal Australian sick berth attendants demonstrating the use of a Neil Robertson stretcher, which constrains patients, enabling them to be transferred from ship to ship at sea, and up and down ladders aboard ship. Right: a Stokes litter used by the RAN for the carriage of sick and wounded personnel.
Australian War Memorial photographs 122534 and 123253

This challenge was addressed differently on each ship, and in some instances considerable structural alterations were needed. Broad approaches, minimum use of stairs, which must be of easy gradient, passenger lifts to the various decks, and careful planning were necessary to ensure an uninterrupted flow of traffic. It was repeatedly found that delays in embarkation of sick and wounded were far more likely to be due to slow movement of the patients through the ship than to tardy delivery of stretchers on the deck.

Finally, the possibility of rapid abandonment of a hospital ship had to be considered: apart from deliberate hostile action a ship might strike a mine or be involved in a marine accident. The lifeboats provided on liners were insufficient in number, for many patients could not help themselves, and they took longer to load, and occupied more space than people who were fully mobile. Up to 50 percent increase on normal requirements for rescue gear was not excessive on a hospital ship; this could be supplied by extra lifeboats, but rafts and Carley floats were found preferable.

Photo 1-7

Carley float on board the cruiser HMAS *Canberra*.
Australian War Memorial photograph P02550.017

CONVERSION OF HOSPITAL SHIP *MAUNGANUI*

Photo 1-8

Hospital ship HMNZHS *Maunganui* at anchor off Port Tewfik, Egypt, May 1941.
National Library of New Zealand

His Majesty's New Zealand hospital ship *Maunganui* was a 30-year-old oil burner with a speed of 15 knots and 7,527 gross register tons. Plans were made for some 390 patients to be accommodated in 100 swinging cots, 100 single fixed cots and 95 double (two-tier) fixed cots. (The number of cots ultimately provided was 365: 22 fracture cots, 84 single cots and the rest two-tier cots.)

The conversion involved a good deal of reconstruction which was carried out by the Wellington Patent Slip Company. In addition, a lot of special equipment was installed. Plans were developed at a series of conferences between the Army Director General Medical Services, Army Headquarters and other Government departments, and the Union Steam Ship Company. The accommodation in the ship was completely stripped and redesigned in the most serviceable manner.

Complete with emergency power, an emergency system of lighting was installed, as well as electric lifts large enough to convey two stretchers from deck to deck. A huge 700-ton fresh water tank was built to ensure adequate water supply between ports.

B Deck was the operating block. This wing contained everything necessary for the equivalent department in a modern hospital. Operating theatre and rooms for sterilising, massage, X-ray, heat therapy, and incidental purposes occupied the space where once was the music room, and nearby was a fully equipped dental surgery. Further aft on the same deck were recreation rooms for officers, men, and nurses, while near the stern the plant was installed for a complete laundry, with modern drying rooms attached.

Photo 1-9

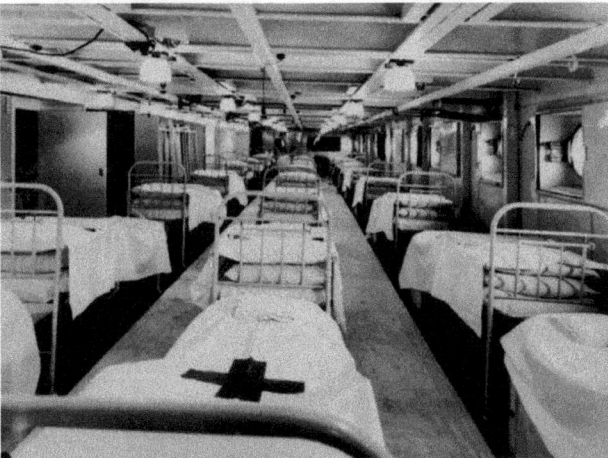

Interior of one of the wards with rows of neatly made beds.
Australian War Memorial photograph 008035

The surgery theatre block consisted of two main units: a plaster room for the casting of broken bones and the theatre proper. Both were exceptionally well appointed when compared with civilian general hospital standards and were equal to the many demands that the *Maunganui*'s varied roles made upon this important section of the surgical side. Not the least of these advantages was the proximity of the X-ray department, which greatly facilitated any procedures requiring the assistance of X-ray screening and radiography. In this connection especially, the orthopedic work benefited greatly.

The plaster room was fully equipped for the purpose – Hawley table, metal sinks and benches for the making of plaster slabs, X-ray viewing boxes, plaster bandage machine and ample cupboard and shelf space. The room was of generous proportions, enabling it to be used as an emergency theatre.

C Deck was devoted mainly to wards.

Photo 1-10

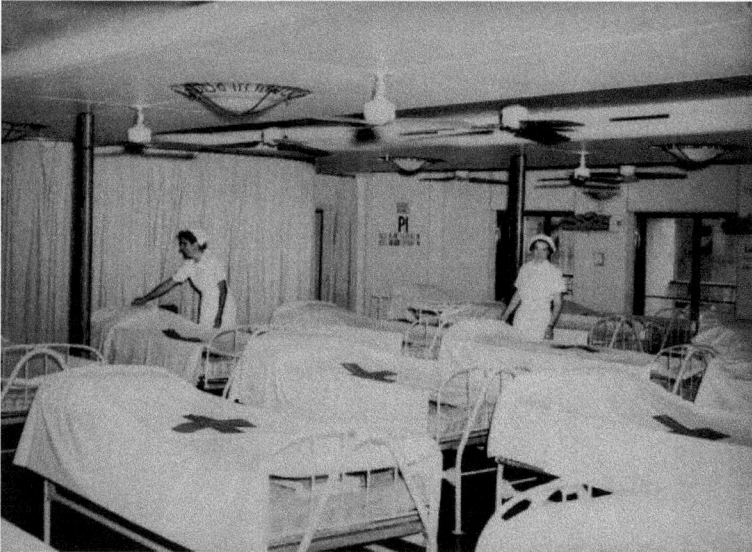

Two unidentified nurses in one of the wards aboard the hospital ship *Oranje*. Australian War Memorial photograph 069583

The operating theatre proper was even larger. Notable features of this important section included a modern operating table with complete accessories and a powerful theatre lamp on an overhead rail, which prevented the lamp swinging to the ship's movement. Adequately sized sterilisers provided hot and cold sterile water. There was even an artificial lung. Numerous extras such as 'angle-poise' lamps were in evidence and a large glass-fronted cupboard provided ample storage for instruments. Properly equipped washbasins were installed. The theatre and the plaster room were finished in a pleasing and restful shade of blue.

Anesthetic equipment again was more than ample. In addition to the usual bottles and masks for general inhalation anesthesia, the theatre unit also had a fully equipped McKesson gas machine and an Oxford vaporiser. Three large side rooms opened off the main theatre and plaster room, while adjacent to both was the theatre storeroom. There was also a surgeons' change-room equipped with shower and toilet, together with locker facilities.

All essential lighting was duplicated on emergency circuits and the whole theatre block was ideally situated forward under the bridge on the promenade deck. The theatre was readily accessible to the main surgical wards owing to its central position and its proximity to the cot-lift.

One feature in which the *Maunganui* differed from the conventional British hospital ship layout was in the siting of the autoclave. This equipment is usually placed in the theatre block, but in the *Maunganui* it was sited some distance aft on the same deck. This was undoubtedly a tremendous advantage in the tropics and prevented the overheating of the theatre. The various specialist departments were all grouped together: theatre, laboratory, X-ray, dispensary and physiotherapy. In addition, all the main cot wards with one exception opened off this central area. This centralisation greatly aided the working of the hospital side of the ship, thus saving time and space.

The main dining room was the main ward. Each of the eight wards had a different colour scheme, and where the lighting arrangements were changed, diffused lighting was installed over the beds.

At the extreme stern of the ship and on the open deck was the isolation ward, complete in itself and entirely separate from the remainder.

Not only was the deck space utilised economically in the provision of wards and incidental accommodation, the holds which once carried cargo and luggage were also converted into quarters of various kinds. The space previously used for No. 4 hold now provided accomodations for NCOs (non-commissioned officers) and orderlies, while another hold became a messroom for the men.

While the power unit of the vessel itself was not changed, a complete emergency system of lighting was fitted, and special conveyors were arranged for the transfer of food from the preparation area to various parts of the ship.

Photo 1-11

Wounded soldiers being transferred from a barge
to the hospital ship *Wanganella*.
Australian War Memorial photograph P10769.001

IN RETROSPECT

As regards the ship and fittings, it was felt that the *Maunganui* was very
suitable for the purpose – she had adequate speed (averaging some 14
knots), stability in heavy seas, well-planned interior hospital
arrangements, and equipment of a high standard.   One standing
criticism was the lack of air conditioning, especially in a ship having to
go through the tropics.

Owing to a variation in requirements with every voyage it was
recommended that a hospital ship should have six or more small rooms
whose special purpose could be determined during each voyage, e.g., for
plaster room, laboratory, special patients. It was felt that a reasonable
amount of deck space for recreation was always necessary, as also was
ample dining space for convalescent patients.

The most important matter in the administration of a hospital ship was the need for harmony between the Army and the Merchant Navy. This centred on the careful choice of the Master and the Commanding Officer Troops. On the *Maunganui* successive Commanding Officers Troops worked in the utmost harmony with the Master, who showed the Army medical staff every consideration.

Once completed *Maunganui* had a crew of 104 medical officers, nurses and orderlies in addition to the ship's crew. She set sail for Suez a day after finishing conversion, leaving on 22 April 1941, arriving on 22 May. She took casualties from Greece and Crete before returning to New Zealand on 10 June 1941. This was a voyage she completed a total of 14 times.

On her 15th voyage, the ship was diverted to the Pacific to assist around various islands making repeated voyages to pick up and treat patients until the end of the war when she brought back one final load of New Zealand Army casualties from Italy and Egypt. By the end of the war the total number of patients she had taken numbered 5,677.

Photo 1-12

Two unidentified Australian soldiers and a Dutch nurse, H. A. Schoonheyt, leaving Port Tewfik onboard *Oranje*, 1941. Australian War Memorial photograph P03836.002

2

Norwegian Hospital Ship
Dronning Maud Sunk

World War II was the most destructive conflict in human history. Tensions in Europe, building for years as a result of aggressive expansion by Fascist Italy and Nazi Germany, culminated in the German invasion of Poland on 1 September 1939. A Nazi-Soviet Non-Aggression Pact signed nine days earlier on 23 August, along with a secret protocol, had left Adolf Hitler free to attack Poland without risking war with the Soviet Union and divided eastern Europe into German and Soviet spheres of influence.[1]

In response to Hitler's pact, both Britain and France had entered into a formal military alliance with Poland. Two days after the invasion of Poland on 3 September, Britain and France declared war on Germany. Britain reluctantly went to war to confront German aggression (which threatened Britain's security and the security of its empire), and defend the balance of power in Europe with force. British leaders strongly believed that German domination of Europe would pose grave risks to the island nation's status and survival.[2]

Seven months later, in April 1940, poorly armed, neutral Norway became the first victim of Germany's Blitzkrieg ("lightning war") in western Europe. Both the Allies and Germany ignored Norwegian neutrality owing to its strategic importance. Because Germany had imported Swedish iron ore that winter through the Norwegian port of Narvik, Britain planned to lay mines along the Norwegian coast. Also, British sailors had boarded the German oil tanker and supply ship *Altmark* in Norwegian waters in February 1940.[3]

Aboard the *Altmark* were sailors taken prisoner and transferred to her from British ships sunk by the pocket battleship *Admiral Graf Spee* while functioning as a commerce raider in the South Atlantic. After *Admiral Graf Spee* was heavily damaged by three cruisers of the Royal Navy's South American Division in the Battle of the River Plate and subsequently scuttled by her crew, in the Río de la Plata in December 1939, *Altmark* had attempted to return to Germany, steaming around the north of Great Britain and then through Norwegian waters.[4]

On 16 February, as the destroyer HMS *Cossack* led other RN ships in pursuit, the *Altmark* ran aground in the shallow waters of Josing Fjord, Norway. The British promptly boarded her and, in the ensuing hand-to-hand combat, overwhelmed the *Altmark*'s crew. The boarding crew searched the German ship and found and freed 299 British merchant ship sailors that were held as prisoners.[5]

Although the Royal Navy ships left Norwegian waters afterward, the *Altmark* incident sowed doubts about respecting Norwegian neutrality among the Allies, as well as the Germans. Both belligerents had contingency plans for military action against Norway. Germany primarily wanted to protect continued shipping of Swedish iron ore (on which the German armaments industry depended in the early stages of the war). Britain wanted to completely stop enemy receipt of the ore.[6]

On 9 April, Germany launched a full-scale invasion of Norway. In a series of surprise attacks, 10,000 German troops seized the capital city, Oslo, and the main sea ports. On land, the poorly equipped Allied troops (Norwegian, British, French, and Polish) were outmatched, and by 2 May most had been evacuated. Fighting continued at Narvik between German and Allied forces until after the Germans had invaded France and Belgium, when it became urgent to save the remaining 24,000 Allied soldiers for use elsewhere. By 8 June, after destroying rail lines and port facilities, they were withdrawn.[7]

Photo 2-1

German troops resting during a break in their march to northern Norway, for invasion of the Scandinavian country in 1940.
Naval History and Heritage Command photograph #NH 71356

DRONNING MAUD SUNK BY ENEMY AIRCRAFT

The last troops had been taken off at Aandalsnes during the night of 30 April/1 May, all the ships from here reaching the UK safely. On [the] 2nd nine He 111s of II/LG 1 made attacks in Ketten strength (three aircraft) against the harbour and town perimeter of Stadlandet, and also against unloading areas at Salangen and Laberget, refuelling at Trondheim en route. One crew attacked the Norwegian passenger vessel Dronning Maud, *and damaged her so badly that she had to be abandoned. This vessel was serving with the Norwegian Army Medical Corps and carried prominent Red Cross markings; casualties aboard were high.*

—From Christopher Shores' book *Fledgling Eagles:*
The Complete Account of Air Operations During the
'Phoney War' and Norwegian Campaign, 1940.

We sat in the lounge and sensed peace in spite of the war – and no danger. He sat at the piano and played, the rest of us sat around… We were soon to go ashore and be sent to Narvik - the front… The ship was clearly marked [with] the Geneva Cross.

Suddenly a salvo came from the side, the windows were splintered and our dear apothecary [pharmacist] fell dead from the piano stool. The ship…then got a couple bomb hits… I managed to get out and let myself down in the sea…, and many of us went swimming towards the dock in Gratangsbotnen. The unbelievable happened—the planes circled over us and shot people who fought for life in the water. Those of us who survived saw people pouring down to their small boats and row out towards us without thinking about the planes.

—Account by a HS *Dronning Maud* (HS Queen Maud) survivor.[8]

The 235-foot Norwegian passenger/cargo ship SS *Dronning Maud* was the first Allied hospital ship sunk in the war. Built in 1925 with the capability to carry 400 passengers, she plied Norwegian coastal waters between Bergen in the south and Kirkenes in the north providing express service. When the Germans invaded Norway on 9 April 1940, the Norwegian government requisitioned her as a troop transport. She made two voyages, transporting Norwegian troops from Sør-Varanger to the Tromsø – Narvik area, and to Sagfjorden.[9]

The *Dronning Maud* then received orders to carry 119 Norwegian Army medics, supplies, eight horses, and three trucks from Sørresia to Foldvik, Norway. For this voyage, she would display a large Red Cross flag (approximately 3 metres by 3 metres) above the bridge deck, and fly smaller Red Cross flags from the foretop and a pole on the boat deck aft.   These means of identification were important because she would be unescorted, carry no anti-aircraft guns for self-protection, and retain her existing black-coloured hull paint.[10]

Photo 2-2

Norweigan hospital ship *Dronning Maud* on fire followng the German aircraft attack. Source: foto.digitalarkivet.no; Arkivreferanse: RA/RAFA-2017/E/Ea/L0088/0013; Arkivverket

Map 2-1

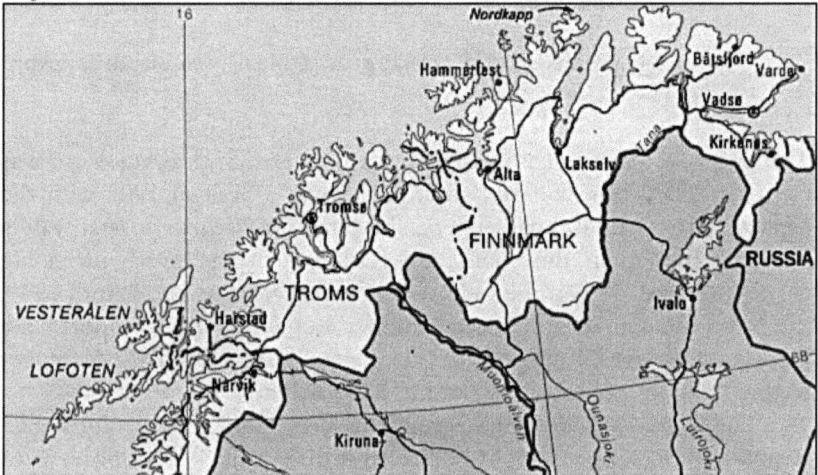

Northern Norway, and adjacent areas of Finland and Russia

Dronning Maud left Sørreisa at 1130 on 1 May 1940, to make the 3-to-4-hour run to Foldvik (not shown on Map 1-1), located about 20 miles north of Narvik on the northwest Norwegian coast (lower left on the map). After entering Gratangen Fjord off Foldvik in mid-afternoon and while manoeuvering to her berth, bombers of Lehrgeschwader 1 arrived. The Luftwaffe (German Air Force) planes immediately attacked the easy target from low altitude, dropping at least seven bombs and raking her with machine-gun fire. Two direct bomb hits—one landing between the stack and the bridge, the other near the after part of the forehatch—blasted large holes in the ship and set her ablaze.[11]

Photo 2-3

German Heinkel He 111 bomber, 9./KG 53 (9th Squadron of Kampfgeschwader/Bomber Wing 53), based at Giebelstadt near Würzburg
Bundesarchiv.
Bild 101I-343-0694-21 / Schödl (e) / CC-BY-SA 3.0

Lehrgeschwader 1 (LG 1) was a Luftwaffe multi-purpose unit in World War II, comprised of fighter, bomber and dive-bomber groups. Formed in July 1936, the unit operated most of the prominent German aircraft utilised during the conflict, such as the Messerschmitt Bf 109, Messerschmitt Bf 110, Dornier Do 17, Heinkel He 111, Junkers Ju 88 and Junkers Ju 87.[12]

It was fortunate that *Dronning Maud* was close to shore because only two of the ship's lifeboats could be launched; the others were either destroyed or inaccessible owing to fire. Most of the survivors apparently made their way ashore by themselves. While still on fire the ship was towed farther offshore and grounded in an effort to save her or facilitate salvage, and then allowed to burn herself out. She later sank, however. Sadly, eighteen people were killed as a result of the attack— eight crewmembers, nine medics, and a tenth medic the following day in the hospital outside Harstad. Additionally, two medical personnel were badly wounded, while twenty-seven were slightly wounded.[13]

The Germans were harshly criticized for sinking this hospital ship by the international press. However, "Queen Maud" had carried out the transportation of soldiers only days earlier. Moreover, although *Dronning Maud* had a Red Cross flag over the bridge deck, and two smaller ones flown aloft, international rules, as explained in the next section, stipulated that hospital ships had to be painted white all over with a green stripe along the entire side of their hulls. The Norwegian ship was painted black, not white.[14]

PEACE CONFERENCE AT THE HAGUE IN 1907

Requirements for hospital ships in wartime had been negotiated and agreed upon decades earlier, at the Second International Peace Conference at The Hague in the Netherlands, in 1907. The Hague Conventions of 1899 and 1907 codified the first multilateral treaties that addressed the conduct of warfare.

Photo 2-4

The Second Hague Conference in 1907.
Gemeente Den Haag www.thehaguepeacejustice.com/peace-and-justice/disclaimer.htm

Hospital ships were covered under the Hague Convention (X) of 8 October 1907, for the Adaptation to Maritime Warfare of the Principles of the Geneva Convention. Most of the stipulations pertaining to hospital ships were covered in the first five articles. For readers not wanting to wade through all the details presented in the following paragraphs, among the most important stipulations were the requirements to:

- Identify to belligerent Powers the names of hospital ships before they were employed
- Distinguish military hospital ships by their being painted white outside with a horizontal band of green about a metre and a half in breadth
- Distinguish hospital ships, equipped wholly or in part at the expense of private individuals or officially recognised relief societies by their being painted white outside with a horizontal band of red about a metre and a half in breadth

The text of the first five articles of the Hague Convention (X) of 8 October 1907, for the Adaptation to Maritime Warfare of the Principles of the Geneva Convention, follow:

Article 1: Military hospital ships, that is to say, ships constructed or assigned by States specially and solely with a view to assisting the wounded, sick and shipwrecked, the names of which have been communicated to the belligerent Powers at the commencement or during the course of hostilities, and in any case before they are employed, shall be respected, and cannot be captured while hostilities last.

Article 2: Hospital ships, equipped wholly or in part at the expense of private individuals or officially recognised relief societies, shall likewise be respected and exempt from capture, if the belligerent Power to whom they belong has given them an official commission and has notified their names to the hostile Power at the commencement of or during hostilities, and in any case before they are employed. These ships must be provided with a certificate from the competent authorities declaring that the vessels have been under their control while fitting out and on final departure.

Article 3: Hospital ships, equipped wholly or in part at the expense of private individuals or officially recognised societies of neutral countries shall be respected and exempt from capture, on condition that they are placed under the control of one of the belligerents, with the previous consent of their own Government and with the authorization of the belligerent himself, and that the latter has notified their names to his adversary at the commencement of or during hostilities, and in any case, before they are employed.

Article 4. The ships mentioned in Articles 1, 2, and 3 shall afford relief and assistance to the wounded, sick, and shipwrecked of the belligerents without distinction of nationality. The Governments undertake not to use these ships for any military purpose. These vessels must in no wise hamper the movements of the combatants. During and after an engagement they will act at their own risk and peril. The belligerents shall have the right to control and search them; they can refuse to help them, order them off, make them take a certain course, and put a commissioner on board; they can even detain them, if important circumstances require it. As far as possible, the belligerents shall enter in the log of the hospital ships the orders which they give them.

Article 5: Military hospital ships shall be distinguished by being painted white outside with a horizontal band of green about a metre and a half in breadth. The ships mentioned in Articles 2 and 3 shall be distinguished by being painted white outside with a horizontal band of red about a metre and a half in breadth. The boats of the ships above mentioned, as also small craft which may be used for hospital work, shall be distinguished by similar painting. All hospital ships shall make themselves known by hoisting, with their national flag, the white flag with a red cross provided by the Geneva Convention, and further, if they belong to a neutral State, by flying at the mainmast the national flag of the belligerent under whose control they are placed. Hospital ships which, in the terms of Article 4, are detained by the enemy must haul down the national flag of the belligerent to whom they belong. The ships and boats above mentioned which wish to ensure by night the freedom from interference to which they are entitled, must, subject to the assent of the belligerent they are accompanying, take the necessary measures to render their special painting sufficiently plain.

As is suggested by the title of this book, and will be gleaned from the text, sanctuary was not guaranteed on board hospital ships. Bitter criticism following enemy attacks on "mercy ships" included wry observations that white paint, red crosses, and lighting of ships at night merely served to easily identify them to belligerent forces bent on their destruction and, accordingly, willing to violate international law.

3

Luftwaffe Destruction of
HMHS *Maid of Kent* and *Brighton*

During World War II, forty-two British hospital ships and hospital carriers served the British Army in evacuating the wounded and injured back to the UK. (The shading in the following table identifies eight hospital ships of other Commonwealth countries. A more detailed table, which denotes which of the British ships were hospital ships and which were hospital carriers may be found in Appendix A.) The latter vessels had shallower draughts and thus could go closer inshore to facilitate swifter evacuation of casualties. They were assisted by flat bottom water ambulances (boats) that could land ashore and take off the wounded in stretchers. In addition to providing soldiers medical or surgical treatment, hospital ships and hospital carriers also offered these services, as required, to Royal Navy and Royal Air Force personnel.[1]

Commonwealth Hospital Ships / Hospital Carriers in World War II

HMHS *Aba*	HMHS *Isle of Jersey*	HMAHS *Oranje*
HMHS *Amarapoora*	HMHS *Isle of Thanet*	HMHS *Oxfordshire*
HMHS *Amsterdam*	HMHS *Karapara*	HMHS *Paris*
HMHS *Atlantis*	HMHS *Karoa*	HMHS *Prague*
HMHS *Brighton*	HMHS *Lady Connaught*	HMHS *Ramb IV*
HMHS *Cap St Jacques*	HMCHS *Lady Nelson*	HMHS *St Andrew*
HMAHS *Centaur*	HMHS *Leinster*	HMHS *St David*
HMHS *Chantilly*	HMCHS *Letitia*	HMHS *St Julien*
HMHS *Dinard*	HMHS *Llandovery Castle*	HMHS *Tairea*
HMHS *Dorsetshire*	HMHS *Maid of Kent*	HMHS *Talamba*
HMHS *Duke of Argyll*	RFA *Maine*	HMHS *Takliwa*
HMHS *Duke of Lancaster*	HMAHS *Manunda*	HMHS *Tjitjalengaka*
HMHS *Duke of Rothesay*	HMNZHS *Maunganui*	HMHS *Vasna*
HMHS *El Nil*	HMHS *Naushon*	HMHS *Vita*
HMHS *Empire Clyde*	HMHS *Newfoundland*	HMAHS *Wanganella*
HMHS *Gerusalemme*	HMHS *Ophir*	HMHS *Worthing*
HMHS *Isle of Guernsey*		

HMHS: His Majesty's Hospital Ship   RFA: Royal Fleet Auxiliary
HMAHS: His Majesty's Australian Hospital Ship
HMCHS: His Majesty's Canadian Hospital Ship
HMNZHS: His Majesty's New Zealand Hospital Ship

Nursing care aboard the floating hospitals was provided by members of the QAIMNS (Queen Alexandra's Imperial Military Nursing Service). These women, like their predecessors in World War I, faced the dangers of war in hostile waters and when carrying out their duties ashore in evacuation ports.[2]

Queen Alexandra's Imperial Military Nursing Service (QAIMNS) was formed in 1902, by Royal Warrant, with Her Majesty as president until her death in 1925. During World War I, the QAIMNS provided nurses across the world, with a Territorial Force Nursing Service being formed, including a branch in India. The size of the QAIMNS expanded in World War II in order to provide nurses and sisters at all Military General Hospitals, and nurses served in all the theatres of operations where British and Indian troops were deployed.[3]

Photo 3-1

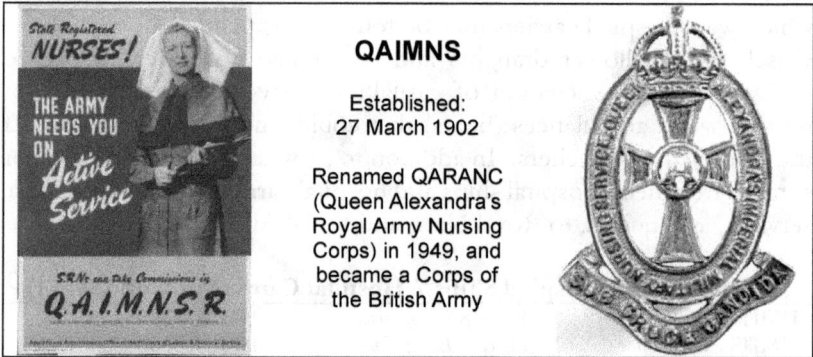

QAIMNS

Established:
27 March 1902

Renamed QARANC
(Queen Alexandra's
Royal Army Nursing
Corps) in 1949, and
became a Corps of
the British Army

WWII British Army Medical Service recruiting poster, and QAIMNS badge.

The QAIMNS became in effect an officers-only unit, because the lowest rank, Staff Nurse, was phased out by 1944. The other ranks from the lowest, Sister, to the highest, Matron-in-Chief, and their Army equivalents were:

QAIMNS Ranks (less Staff Nurse) in World War II

Sister – Lieutenant	Principal Matron – Lieutenant Colonel
Senior Sister – Captain	Chief Principal Matron – Colonel
Matron – Major	Matron-in-Chief – Brigadier[4]

Members of the QAIMNS, although treated as officers, were not commissioned officers of the British Army until 1949. This action coincided with the service being admitted as a corps of the British Army, and redesignated as the Queen Alexandra's Royal Army Nursing Corps.[5]

In WWII, International Law of the Geneva Convention stipulated that hospital ships had to display electrically powered lights to illuminate their red cross signs on the sides of the ship and upon the deck. The ships were painted all white with a broad green or red stripe around the hull and red crosses painted on the sides to make them easily identified. These distinguishing markings and lighting were intended to identify to enemy aircraft, U-boats, and warships that they were not legitimate targets. Sadly, the reality was that they pinpointed easy targets.[6]

HOSPITAL CARRIER *MAID OF KENT*

Photo 3-2

British hospital carrier *Maid of Kent*, circa 1939-1940.
British Government photograph

The 329-foot passenger ferry SS *Maid of Kent* was built in 1925 by William Denny & Bros, Dumbarton Leven Yard, located on the Leven River near its junction with the River Clyde. Naming of the ferry for the fourteenth century nun Elizabeth Barton, later known as the Maid of Kent, is interesting based on her life and its abrupt end. Born about 1506, Barton, who started off life as a humble servant girl, who later became both a mystic, and a nun, was one of the few women who dared to question the King Henry VIII's judgment. Famous for speaking her mind to powerful people, this habit eventually proved to be her undoing when she was executed for her outspokenness.[7]

The Maid of Kent moniker results from Barton, a poor girl, working in service for Thomas Cobb in Adlington, in Kent, who apparently had visions and communications with godly presences. Cobb had been a bailiff and steward and helped run the Archbishop of Canterbury's estates in the parish. He asked the priest, Richard Masters, to visit her, along with a local man, Edward Thwaites. Following this visit, they and many others, who were amazed at her insights, started to think of her as a prophetess.[8]

Masters wrote a report about her to the Archbishop of Canterbury, William Warham, who later, after Elizabeth said she wanted to become a nun, arranged for her to enter a convent, the Benedictine House of St. Sepulchre's in Canterburyhis—a 400-year-old nunnery.[9]

Years passed and, as Elizabeth continued to share her visions, both her reputation and her influence on the Church and the King's court grew. Her downfall started in 1528, when King Henry VIII began trying to get a divorce from Catherine of Aragon, his first wife who was Queen of England from their marriage on 11 June 1509 until its annulment on 23 May 1533. Elizabeth met with Henry VIII several times, and each time spoke against his planned marriage to Anne Boleyn following the divorce he was seeking.[10]

Elizabeth told him that she heard threats from God that Henry would not live 7 months if he married Anne Boleyn, and she told others that he would not be king a month after marrying Boleyn. Elizabeth even corresponded with the Pope about the matter writing that there would be plagues if the Pope favored Anne Boleyn, and that the Pope himself would be destroyed if he didn't support Catherine of Aragon.[11]

Sir Thomas More, who served Henry VIII as Lord High Chancellor of England, met with Elizabeth and later wrote her that it was his advice that she should stop talking about the King and the divorce.[12]

Henry married Anne in secret. When he didn't die, people began to doubt her prophesies, and Henry himself became furious at the nun. Henry asked his minister Thomas Cromwell to have Thomas Cranmer, who had replaced Warham as Archbishop of Canterbury, begin an investigation. By autumn 1533, Cromwell had rounded up everyone who had known and supported her, and had them arrested, and all pamphlets and books found about her and her prophecies were burned.[13]

In November 1533, Elizabeth Barton had to do a public penance by reading a confession that she was, "a most miserable and wretched person, who had been the origin of all this mischief, and by my falsehood have deceived all these persons here and many now present." This forced act of contrition was apparently not sufficient. Parliament sentenced her to death, individuals could be condemned without a trial, and on 20 April 1534 she was hanged. Her final words were:

> Hither am I come to die, and I have not been the only cause of mine own death which most justly I have deserved, but also I am the cause of all these persons which at this time there suffer. And yet to say the truth, I am not so much to be blamed considering it was well known unto these learned men that I was a poor wench without learning and therefore they might have easily perceived that the things that were done by me could not proceed in no such sort. Because the things which I feigned was profitable unto them, therefore they much praised me and bare me in hand that it was the Holy Ghost and not I that did them, and then I being puffed up with their praises fell into a certain pride and foolish fantasy with myself and I thought I might feign what I would, which thing hath brought me to this case.[14]

Barton's head was chopped off to be mounted at London Bridge, but her headless body was permitted burial in the church of the Grey Friars in London.[15]

GERMAN INVASION OF WESTERN EUROPE

On 10 May 1940, Germany began a campaign against the Low Countries and France, which lasted less than six weeks as German troops overran Belgium, the Netherlands, Luxembourg, and France. This rapid victory was facilitated in part by British and French commanders erring in believing that German forces would attack through central Belgium as they had in World War I, and rushing forces to the Franco-Belgian border to meet the German attack.[16]

The main German attack, however, pressed forward through the Ardennes Forest in southeastern Belgium and northern Luxembourg. German tanks and infantry quickly broke through the French defensive lines and advanced to the coast. The Netherlands surrendered on 15 May, and Belgium on the 28th. More than 300,000 French and British troops were miraculously evacuated from the beaches near Dunkirk, France, across the English Channel to Great Britain.[17]

Paris, the French capital, fell to the Germans on 14 June, and eight days later, France signed an armistice on the 22nd. This left Britain as the only country fighting Nazi Germany. As part of the armistice agreement with France, Germany occupied northern France and all of France's Atlantic coastline down to the border with Spain.[18]

MAID OF KENT AT DIEPPE

In September 1939, the *Maid of Kent* and her sister ship, the *Isle of Thanet* were requisitioned and converted to hospital carriers. In addition to interior modifications, they were painted white with a green band running completely around them. Predominantly displayed on both sides of their hulls were two large red crosses and their funnels (stacks) likewise had red crosses on either side. Finally, bright green lights all around them clearly illuminated and identified them as non-combatant hospital ships.[19]

The *Maid of Kent* and *Isle of Thanet*, both built by William Denny and Bros. Ltd. of Dunbarton for the Southern Railway Company, were designed for short sea routes across the Channel: Dover to Calais, and Folkestone to Boulogne. *Maid of Kent*, 342 feet long and 45 feet wide and (powered by five Babcock-Wilcox water-tube boilers sending steam to four Parsons turbines, turning twin screws via reduction gears) could make a very respectable 22 knots.[20]

In early evening on 18 May, 1940 *Maid of Kent* and another hospital ship carrier *Brighton* sailed from Newhaven, a port town in southeastern England, bound cross-channel for Dieppe, a fishing port on the Normandy coast of northern France. In 1940, Newhaven was the main base for British hospital ships operating between England and France. A number of the ships employed besides the *Maid of Kent* were also converted cross-channel steamers, these included the hospital carriers *Brighton*, *Isle of Thanet*, and *Paris*.

Map 3-1

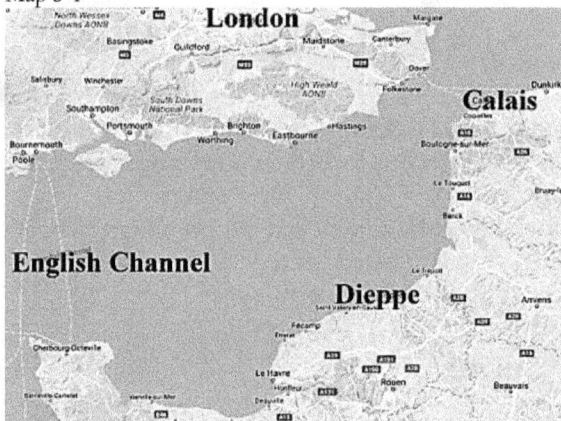

English Channel area; Newhaven lies about midway between
Brighton and Eastbourne; Dunkirk is located to the east of Calais

Maid of Kent (Captain Leonard Addenbrooke) and *Brighton* had been ordered to Dieppe to pick up wounded servicemen being evacuated from France. *Maid of Kent* arrived at 2120, about the same time *Brighton* also entered port. Thirty minutes later the harbour was attacked by the Luftwaffe in a well-planned bombing raid, but neither ship was harmed.[21]

Maid of Kent's avoidance of attack on 18 May was probably related to her initial berthing within a gated dock. It probably helped that a large area of lawn beside the quay had been covered with white chalk and a large red cross had been sprayed on. (A quay or quay wall is part of the river bank or coastline which has been modified so ships can dock at it parallel to the shore.) Aboard the hospital carrier, a large white canvas was stretched between the mainmast and the stern flag post, with a large red cross painted on. However, as the danger of air attack heightened the captains of both ships requested permission to move their vessels from the harbour inner basin to the tidal Maritime Station quay, for ease of departure in an emergency.[22]

On 21 May, permission for the vessels to shift berths was finally granted. *Brighton* moved toward the lock gates but the swing footbridge, jammed, leaving both hospital carriers in the inner basin. Later that day, a train pulled up alongside *Maid of Kent*, and stretcher cases were taken off and laid on the quayside, waiting to be taken aboard her.[23]

At 1700, the Luftwaffe attacked again. The first wave of bombers attacked the harbour but missed the *Maid of Kent*. However, the second wave of bombers swept in, and she was hit. One bomb dropped straight down the funnel into the engine-room, another two hit the afterdeck, and the ship was completely ablaze within a few minutes. Seventeen of the crew were lost, along with eleven RAMC (Royal Army Medical Corps) personnel.[24]

Photo 3-3

Royal Army Medical Corps (RAMC)

Established in 1898, when the Army Medical Staff merged with the Medical Staff Corps

The RAMC continues to attend to the health of servicemen and women to this day

Royal Army Medical Corps (RAMC) World War I poster, and a RAMC cap badge.

Decades later, at age 88, William (Bill) Warman described being up on deck with Fred (Bisto) Pilcher, a cook from the officers and passengers galley, and watching the second wave of bombers come in and drop their bombs:

> We judged where they would land and ran forward as fast as we could. We ran below decks to the first class passenger lounge below water level. A bomb dropped straight down the funnel, another through the engine room skylight. Two fell onto the afterdeck, one of them ploughing directly into the crews' galley, yet another fell between the quayside and the ships portside. When the water began to rush in, we ran back up and on to the quay. The Hospital Ship was completely ablaze within three to four minutes. The mainmast had fallen and at least one engine had exploded. The people below decks didn't stand a chance.[25]

The *Maid of Kent* was completely destroyed and sank. She was later raised by the Germans and abandoned in deeper water outside the confines of the harbour.[26]

The train was strafed by the aircraft and set aflame, but none of the injured soldiers on the quayside were further hurt. Warman credited a locomotive crew with saving many lives:

> Those stretcher cases and the walking wounded had been removed from the quay at the first sign of the enemy raid, none were further injured. The front half of the train had the bulk of the stretcher cases. A French Shunter [driver of a shifter locomotive used for manoeuvering railway vehicles over short distances], who I do not think was ever recognised for his bravery, uncoupled the train about halfway along and then moved the front half, inland – away from the attack. He saved their lives. The rest of the train, which had been strafed by German aircraft, was burned out.[27]

HOSPITAL CARRIER *BRIGHTON* ALSO SUNK

During the enemy air attack that destroyed *Maid of Kent*, *Brighton* was abandoned of all crew, medical staff and patients. She was, however, undamaged in the air attack, so it was decided to muster a crew to sail her but the only engineers to be found were survivors from *Maid of Kent*, who were unfamiliar with *Brighton*'s boilers. Details are scarce and differ regarding the details and day of her ensuing loss. A British army officer who was at Dieppe at the time, stated that she was sunk the evening of 22 May while alongside the Greek ship *Galaxias*. Her owners recorded the date of her loss as 24 May, as also did Admiralty and Lloyds of London reports.[28]

4

British Hospital Carrier *Paris* Sunk at Dunkirk

I feel as Captain of this ship, and now I have a few moments to spare, that I should like to give expression to my admiration and deep regard for the nursing sisters of this ship.

We have recently made two trips to Dunkirk and two to Cherbourg, in each case being the last Hospital Ship to enter and leave the ports. Our second trip to Dunkirk was under extremely severe conditions, bombs and shells dropping all about us and men being wounded and killed alongside our ship on the pier. We had numerous narrow escapes and nerve racking experiences.

During all of this our Nurses were really splendid; never a sign of excitement nor panic of any kind; they just carried on, under the able leadership of our Matron calmly and efficiently, and I feel quite sure that their magnificent behavior was an important factor in steadying up the members of the RAMC [Royal Army Medical Corps] personnel with whom they worked.

My sentiments are warmly endorsed by every member of the crew of this ship.

—Letter from Captain John Ailwyn Jones to the Matron-in-Chief
of the Queen Alexandra's Imperial Military Nursing Service
at the War Office in London, expressing his admiration
for the service of Matron Thomlinson and her nurses
on board the British hospital ship *Dinard*.[1]

On 31 May 1940, Dunkirk was ablaze as the hospital carrier *Isle of Guernsey* berthed alongside the quay. She had been waiting outside the wreck-strewn harbour for four hours, before loading over 600 wounded men on a ship with space for only 203 cot cases. Most were passed over the side of the ship, because gangplanks could not be used to board or leave the vessel at this badly damaged area of the port.[2]

Isle of Guernsey made five voyages in just over two weeks to evacuate soldiers/patients from Cherbourg, Boulogne and Dunkirk. Following return of the ship to Poole, a seaport on the south coast of England,

Sister Dora Grayson in charge of the Queen Alexandra's nurses on board the hospital carrier recalled this about her voyages to Dunkirk:

> To Dunkirk...the German plane returned and dived backwards and forwards dropping (they said) 10 salvoes of 3 bombs each & machine-gunning all the time. One cannonball went straight through the foremast at the level of the Bridge, which proved that the pilot must have been low enough to see all 5 large Red Crosses...Just as all had given up all hope of survival, an RAF plane came and drove off the Germans.[3]

OPERATION DYNAMO

Sister Grayson's reference to Dunkirk, France, concerned the Dunkirk evacuation which took place between 26 May and 4 June 1940. Codenamed Operation DYNAMO, the seemingly impossible, but highly successful evacuation was subsequently commonly referred to in the press as the "Miracle of Dunkirk," or just Dunkirk for short. Vice Adm. Sir Bertram Ramsay, flag officer Dover Command who oversaw the operation, was similarly dubbed the "Savior of Dunkirk."

Photo 4-1

Drawing by Dwight C. Shepler of British Admiral Sir Bertram H. Ramsay, RN, who engineered the Dunkirk evacuation, and later commanded the British Task Force at Sicily, and the great Allied amphibious force of continental invasion at D-Day, 1944. Naval History and Heritage Command photograph #88-199-EK

In May 1940, as German forces began inflicting a series of defeats on the Allies in the Low Countries (Belgium, the Netherlands, and Luxembourg) and France, Ramsay was approached by Winston Churchill to begin planning an evacuation. (Churchill, who had been appointed First Lord of the Admiralty at the outbreak of the war, succeeded Neville Chamberlain as prime minister on 10 May.) Meeting in Ramsay's headquarters in tunnels beneath Dover Castle, the two men planned a large-scale evacuation of the British Expeditionary Force (BEF) at Dunkirk. Pushed up against the sea by a Germany Army, the only hope of escape for these troops, along with French and Belgian soldiers, was by sea across the English Channel.[4]

Initially hoping to evacuate 45,000 men over two days, Ramsay employed a massive fleet of disparate vessels which ultimately saved 338,682 British, French, Belgian and Dutch personnel over nine days. Ramsay's destroyers brought back 103,399, followed by merchant ships, 74,380, and minesweepers, 31,040; while a huge variety of other vessels including fishing vessels and private yachts, the rest. Approximately 850 vessels of all shapes and sizes took part, of which nearly 240 were lost.[5]

Some readers may envision an armada of "Little Ships" retrieving troops from the beaches at Dunkirk, perhaps owing to a popular 2017 epic historical war thriller film titled Dunkirk. In reality, two-thirds of those rescued was via the Eastern Mole (pier) and Quay at Dunkirk harbour by Royal Navy destroyers, large troop transport ships, ferries, hospital ships, and minesweepers. Six Royal Navy destroyers were sunk in these operations and a number of the transports and ferries and the hospital carrier *Paris* were also lost.[6]

However, the role of the civilian craft in rapidly evacuating trapped soldiers was highly important. Time and space limitations did not allow the evacuation of all the troops via the harbour and hence, evacuation from the beaches was required. With the beaches at Dunkirk shelving so gently, small craft were able to rescue men either directly, or to ferry them to larger vessels waiting offshore. Among the ranks of this heroic collection of small vessels sourced from southeast England, were river launches, old sailing and rowing RNLI (Royal National Lifeboat Institution) lifeboats, yachts, pleasure steamers, fishing boats, commercial sailing barges, and Thames fire boats.[7]

Not only did Royal Navy ships and civilian vessels have to run the gauntlet of round-the-clock air attacks during daylight hours, they were also when inshore, within the range of German artillery. Offshore, they faced the ever-present threat of German mines and U-Boats and E-Boats of the German Kriegsmarine (Navy), not to mention the natural perils of the swells and currents of the English Channel.[8]

On 4 June 1940, Prime Minister Winston Churchill described in a speech before the House of Commons, the all-out efforts German forces had mounted in an attempt to destroy the vessels carrying soldiers across the English Channel to safety in England. Two paragraphs of his speech follow; the last sentence honours the dangerous duty of the hospital ships that took part in the operation:

> The enemy attacked on all sides with great strength and fierceness, and their main power, the power of their far more numerous Air Force, was thrown into the battle or else concentrated upon Dunkirk and the beaches. Pressing in upon the narrow exit, both from the east and from the west, the enemy began to fire with cannon upon the beaches by which alone the shipping could approach or depart. They sowed magnetic mines in the channels and seas; they sent repeated waves of hostile aircraft, sometimes more than a hundred strong in one formation, to cast their bombs upon the single pier that remained, and upon the sand dunes upon which the troops had their eyes for shelter. Their U-boats, one of which was sunk, and their motor launches took their toll of the vast traffic which now began. For four or five days an intense struggle reigned. All their armoured divisions-or what was left of them-together with great masses of infantry and artillery, hurled themselves in vain upon the ever-narrowing, ever-contracting appendix within which the British and French Armies fought.

> Meanwhile, the Royal Navy, with the willing help of countless merchant seamen, strained every nerve to embark the British and Allied troops; 220 light warships and 650 other vessels were engaged. They had to operate upon the difficult coast, often in adverse weather, under an almost ceaseless hail of bombs and an increasing concentration of artillery fire. Nor were the seas, as I have said, themselves free from mines and torpedoes. It was in conditions such as these that our men carried on, with little or no rest, for days and nights on end, making trip after trip across the dangerous waters, bringing with them always men whom they had rescued. The numbers they have brought back are the measure of their devotion and their courage. The hospital ships, which brought off many thousands of British and French wounded, being so plainly marked were a special target for Nazi bombs; but the men and women on board them never faltered in their duty.[9]

HOSPITAL CARRIER *PARIS* LOST

At 2250 on 2 June, Capt. William Tennant, who had tactical oversight of the evacuation and was designated "beachmaster," radioed Admiral Ramsay the triumphant message "BEF evacuated." Tennant and British

I Corps commander Gen. Harold Alexander toured off the beach and in the harbour area in a motor launch, calling out with a megaphone to ensure that no BEF evacuees had been missed.[10]

Three and a half hours earlier, the last ship in passage across the Channel to recover wounded men from the Dunkirk harbour came under attack by a German bomber near the French port. The hospital carrier *Paris* had made her first of several visits to Dunkirk on 25 May to assist with the evacuation. On this final voyage on 2 June, she was initially attacked by a single German bomber at 1915, with resultant damage that disabled her and killed two crewmen. Further attacks by Stukas and additional damage followed. A tug was sent to assist *Paris*, but seawater entering her via hull damage, caused the hospital carrier to gradually settle, and she sank around midnight that day.[11]

A lifeboat carrying nurses was also bombed, and several of the sisters sustained bad injuries. They and other brave women aboard other hospital carriers had sailed back and forth across the English Channel several times, enduring bombs, mines and attacks by the Luftwaffe. Rarely stopping to think about the prevailing danger, their focus was on caring for the wounded men entrusted to their care.[12]

5

Sinking of Five Greek Hospital Ships

ITALIANS BOMB GREEK HOSPITAL SHIP.
LONDON, March 13. (A.A.P.)

An Italian plane attacked the Greek hospital ship Socrates *(3,169 tons) in the Ionian Sea, states the Greek Ministry of Marine.*

The ship, which was loaded with Greek wounded from Albania, was clearly marked, the announcement adds, and the Italians had been notified in advance of the ship's sailing.

Bombs were dropped from only 350 feet, and three of them fell 50 yards from the stern of the vessel, slightly damaging the deck.

—The Sydney Morning Herald, 14 March 1941.[1]

Shortly after 0300 on the morning of 28 October 1940, the fascist Italian government of dictator Benito Mussolini sent an ultimatum to Greece through the Italian ambassador in Athens. Emanuele Grazzi personally delivered this message to Prime Minister Ioannis Metaxas at his home in Kifissia Athens. It demanded the free passage of the Italian army from the Greek-Albanian border in order to then occupy some strategic points in Greece such as seaports and airports for its supply needs and other facilities, incident to its subsequent movement to Africa. After reading the message, Metaxas replied to the Italian ambassador in French (the official diplomatic language) "Alors, c'est la guerre," ("Well, this means war"), thus denying the Italian mandate.[2]

Only hours later, at 0530 that day, the Greco-Italian war began with the surprise invasion of Italian troops into Epirus, a coastal region of northwestern Greece and southern Albania. At this time, Italy occupied Albania, and Mussolini, bolstered by this initial success, had decided that his forces could also prevail in Greece against minimal opposition.[3]

However, the Albania-based Italian army that launched the invasion encountered unexpectedly tenacious resistance by the Hellenic army during combat in the mountainous and muddy terrain on the Albanian–Greek border. By mid-November 1940, the Greeks had stopped the Italian invasion just inside Greek territory. As British aircraft struck Italy's forces and bases, the Greeks pushed the Italians back into Albania. The Greek offensive culminated in the capture of Klisura Pass in southern Albania in January 1941.[4]

Map 5-1

Greece and adjacent areas of Albania, Yugoslavia, and Bulgaria

This was humiliating for Mussolini; what was supposed to have been an easy victory became a desperate fight for survival of his army. The Greco-Italian war, as previously stated, began on 28 October 1940, initiated the Balkans Campaign between the Axis powers and the Allies and eventually turned into the Battle of Greece with British and German involvement.[5]

The front stabilized in February 1941, with 28 Italian divisions opposing the Greeks' 14 divisions (which were larger). In March, the Italians conducted an unsuccessful spring offensive. At this point, losses were mutually costly, but the Greeks had far less ability to replace their losses in men and materiel, and they were dangerously low on ammunition and other supplies. This situation provided an opening for the British to enter Greece and establish an airbase in Athens, which would be within striking distance of valuable oil reserves in Romania, which Hitler relied upon for his war machine.[6]

On 7 March, the first of 58,000 British and Australian troops diverted from Egypt arrived in Greece to occupy the Olympus-Vermion line. The following month, Hitler came to the aid of his Axis ally. On 6 April, the Germans invaded northern Greece at the fortified Greek Metaxas Line, along the Greco-Bulgarian border. The complex of concrete bunkers and fortresses was manned by only a third of its authorised strength—the vast majority of Greek forces being deployed in the mutually costly stalemate with the Italians on the Albanian front. Nevertheless, defence forces inflicted heavy losses on the attackers before being overrun by Lt. Gen. Franz Böhme's XVIII Mountain Corps.[7]

Overwhelming the Greek and British forces in northern Greece, the Germans advanced rapidly west and south. Greece surrendered to German troops on 20 April 1941; three days later, on the 23rd, the Greek Government fled Athens for Crete and the Northern Greek Army surrendered to the Italians.[8]

On 24 April, six Royal Navy cruisers, 24 destroyers and escort ships, 2 infantry landing ships, and 14 troopships took part in the main evacuation of Allied troops from Porto Rafti and Rafina, port cities in east-central Greece bordering the Aegean Sea.[9]

ATTACKS ON GREEK HOSPITAL SHIPS

On 6 April 1941, the Germans invaded the country from the Greek-Bulgarian borders and due to the power of their air force managed to move further south. Following successive air strikes, almost 120 Greek vessels, including the entire passenger fleet and five hospital ships were hit and sunk in Greek waters. Only four passenger vessels managed to survive escaping to the Middle East.

—Summary information concerning the decimation of Greek shipping at the hands of the Luftwaffe in spring 1941.[10]

In March-April 1941 during the Greco-Italian war, the German Luftwaffe and Italian Regia Aeronautica (Royal Italian Air Force) carried out a series of deadly air attacks against Greek hospital ships. These attacks on unarmed vessels clearly marked as hospital ships, demonstrated that neither the Italian Fascist regime nor the Nazi Germans respected the Hague Convention of 1907.[11]

The first air strike was conducted on 12 March 1941 by an Italian bomber (probably an Italian-made CANT Z.1007 Alcione tri-engine bomber), which attacked the Greek hospital ship *Socrates* carrying wounded soldiers from the front, in the Ionian Sea off Lefkada Island. The first large-scale use of these type of bombers had taken place during the Italian invasion of Greece in October 1940, while a few were used in the later stages of the Battle of Britain.[12]

The Italian bomber dropped its ordnance from a height of approximately 350 feet, which fortunately missed the hospital ship. No casualties resulted from the attack and, having received only minimal damage, *Socrates* continued her voyage.[13]

HOSPITAL SHIP *ATTICA* SUNK

Photo 5-1

Two German Junkers Ju 87 "Stuka" dive bombers in flight, Europe, circa 1941. Australian War Memorial photograph 106483

HS *Attica* was the first Greek hospital ship lost to enemy attack. The elderly 281-foot vessel was the former steel-screw steamer *Grenada* built in 1896 by Alexander Stephen & Sons Ltd., Govan, Scotland. On the night of 11 April, she was transiting a strait in the Aegean Sea between the islands of Euboea and Andros when she was attacked at 2330 by German Stuka dive-bombers off Cape Cafireas (or Cavo D' oro), a promontory on the southeastern tip of Euboea, and was sunk.[14]

The Ju 87 Stuka was a single-engine monoplane, easily recognisable by its inverted gull wings, used by the German Luftwaffe with especially deadly effect during the first half of World War II. (Stukas were highly effective against shipping, but no match for British Hurricane and Spitfire fighters when the Luftwaffe did not have control of the skies.) The aircraft was designed to dive on the target at a steep angle and release its bombs at low altitude for maximum accuracy before pulling up. It had dive brakes to slow the dive, giving the pilot more time to position his airplane and, thereby, aim the bombs. Hinged external bomb racks could swing downward and outward so that the bombs, when released, would clear the airplane's propellers.[15]

Stukas were armed with four 7.9-millimetre machine guns, two of which were operated by a rear-seat gunner; the latter guns were replaced late in the war by a single 13-millimetre gun. The most common bomb load out for a Stuka was a single SC250 general-purpose 500-lb bomb carried beneath the fuselage and four SC50 110-lb fragmentation bombs under its wings. Particularly terrifying for those below, was the wail that Stukas made as they swooped down to release their bombs. Fulfilling their intended purpose as a psychological weapon, many soldiers described the sound made by air-driven sirens attached to the planes' wheel struts, as being particularly horrifying.[16]

Equipped with 362 beds, *Attica* had sailed from Kavala (a coastal town in northeastern Greece) and Thásos (northernmost island in the Aegean Sea) with 11 wounded soldiers, 28 nurses, 17 military physicians, and 110 other men comprised of embarked officers and non-commissioned officers, and sailors (presumably members of the crew).[17]

Despite being fully lit and bearing the Red Cross credentials, she came under aerial attack, resulting in loss of the ship and the deaths of twenty-eight people including the ship's master, Captain Dimitrios Meletopoulos. Actions by the *Attica*'s crew and medical staff saved the lives of many passengers.[18]

HOSPITAL SHIP *ESPEROS* SUNK, *ELLENIS* DAMAGED AND SUBSEQUENTLY SCUTTLED

On 21 April 1941, German air and ground forces attacked British, Australian, and New Zealand troops at Thermopylae, Greece. As part of that offensive, dive bombers sank the Greek torpedo boat *Thyella*, and several freighters. The hospital ships *Esperos* and *Ellenis* were also part of their prey.[19]

The hospital ship *Esperos* was bombed and sunk while at anchor off Messolonghi (a port town on the Gulf of Patras in western Greece), taking on wounded. *Esperos* had been built in 1918 as the convoy sloop HMS *Ivy*, by Blyth Shipbuilding & Drydock Co. Ltd., located on the river Blyth in Blyth, Northumberland, England.[20]

German bombers also damaged the Greek hospital ship *Ellenis* on the 21st near Patras. She was brought to Patras and her wounded on board were disembarked. *Ellenis* was scuttled later in the month, and subsequently refloated by the Germans.[21]

GREEK HOSPITAL SHIP *SOKRATIS*

The following day, 22 April, German dive bombers attacked shipping in the Saronic Gulf, and the Gulf of Corinth northwest across the isthmus of Corinth. The Saronic Gulf, formed between the peninsulas of Attica (which includes Athens) and Argolis (of which Nafplio is the capital), defines the eastern side of the isthmus of Corinth and is part of the Aegean Sea. Piraeus, a port city in the Attica region of Greece, lies on the Saronic Gulf (lower centre area of Map 5-2).

Map 5-2

The Corinth canal connects the Saronic Gulf (lower half of the map by Piraeus) to the Gulf of Corinth, which lies to northwest across the isthmus of Corinth

Sunk in these two gulf areas were the Greek minelayer *Aliakmon*, hospital ship *Sokratis*, 11 freighters, and 1 tanker. The *Sokratis* was sunk at Antikyra, a port on the west coast of the Gulf of Antikyra (a north-coast bay of the Gulf of Corinth). In mid-morning on the 22nd, twenty-four Ju 87 Stukas were observed overhead. During the first wave of dive-bombing attacks, *Sokratis* escaped harm.[22]

Her respite was short-lived. In the second wave, approximately 50 bombs were dropped. One hit *Sokratis* between the funnel and stern. The resultant blast breeched her hull, sea water flooded in, and she immediately took on a list. During the third wave of attacking Stukas, three more bombs hit the hospital ship; one on the aft winch and two between the funnel and the after mast. *Sokratis* then swiftly sank stern first; the bombed tankers *Thedol* and *Theodora* also sank.[23]

GREEK HOSPITAL SHIPS *POLICOS* AND *ANDROS*
A fourth and fifth hospital ship, *Policos* and *Andros*, were lost to German bombers on 23 and 24 April. On the 23rd, German bombers destroyed 13 British Hurricane fighters on the ground at Argoes, and sank the Greek torpedo boat *Kios*, minelayer *Nestos*, battleships *Kilkis* and *Lemnos*, 12 freighters, 1 tanker, and the hospital ship *Policos*. The damaged Greek torpedo boat *Doris* was scuttled to prevent her capture.[24]

King George II of Greece, Crown Prince Paul, Prime Minister Emmanuel Tsouderos and other important figures of the Greek government were evacuated by the RAF to Crete, the first stage of an eventual evacuation to Egypt. As an act of appreciation, the king instructed that his wine cellar contents be distributed to the Allied troops who fought to defend his country.[25]

On 24 April, the Greek hospital ship *Andros* was sunk by German bombing off Loutraki, on the Gulf of Corinth coast. The following day, the 25th, German paratroopers captured the Corinth Canal. Meanwhile, German aircraft sank, in addition to the *Andros*, eleven freighters off the coast, and damaged British cruiser HMS *York* and submarine HMS *Rover* at Suda Bay, Crete. Greek torpedo boats *Aigli*, *Alkyoni*, and *Arethousa* were scuttled to prevent their capture.[26]

At 1900 on 24 April, the Greek luxury yacht *Hellas* was bombed at Piraeus while boarding British civilians and wounded Allied soldiers, killing 500. The large steam yacht had arrived unexpectedly in the harbour earlier that day, and her captain offered to take 1,000 passengers. The ship was instructed to depart after dark, and about 500 British civilians (mostly Maltese and Cypriots) and 400 sick and walking wounded were embarked on board. Among them were 75 New

Zealanders of 4 Reserve Mechanical Transport (RMT) Company and a similar number from 28 (Maori) Battalion.[27]

Just before sunset, seven German Stukas attacked the *Hellas*. Five bombs struck the yacht, setting her on fire, while another three burst alongside on the jetty. The only gangway to the jetty was destroyed, and passengers were trapped in burning cabins. There were no working fire hoses on board and none were available ashore for nearly an hour. The vessel rolled over and sank, with the loss of 400 to 500 lives, and survivors—many with terrible wounds—struggled into a nearby warehouse. Commandeered vehicles ferried them to a hospital in Athens, where most were later taken prisoner.[28]

Of the 75 men in 4 RMT Company's Workshops Section, 31 made it to Crete seeking evacuation, but only seven later rejoined the section in Egypt. Most of 28 (Maori) Battalion was evacuated from the Athens area to Crete aboard the large infantry landing ship HMS *Glengyle* in the early hours of Anzac Day (25 April). They left behind 10 dead and 81 prisoners of war.[29]

Photo 5-2

British naval ships in Suda Bay, Crete, in May 1941, during the evacuation of Allied forces. At the foreground, Australian troops are resting near the shores of the bay. Australian War Memorial photograph 073480

6

British Hospital Ship *Ramb IV*

Photo 6-1

Painting by William Dargie of the hospital ship *Ramb IV* at Haifa, 21 January 1942.
Australian War Memorial photograph ART22153

On 4 July 1941, an article titled "31 Willful, Brutal Attacks on Hospital Ships" appeared in the New Zealand newspaper *The Northern Advocate*. Among information provided readers was that a year earlier, the British Government had protested to Germany through the United States on 12 July 1940, about 31 deliberate and flagrant attacks by aircraft and shore batteries upon British hospital ships and carriers. The sinkings of *Maid of Kent*, *Brighton*, and *Paris*, and at least half of the other attacks had occurred in broad daylight. Moreover, all the ships were marked in accordance with the Hague Convention, and in all respects conformed with other requirements of that convention.[1]

In spite of this protest, enemy aircraft continued to attack British hospital ships/carriers. One example, among many, was the hospital ship *Vita*. With over 400 casualties aboard, she was deliberately bombed by an enemy aircraft on 12 April 1941, and attacked again on the 21st and 22nd while at sea and under tow. (Summary information about the dozens of British hospital ships/carriers attacked by German and Italian (mostly air) forces between 18 May 1940 and 10 May 1942 may be found in Appendix C.)[2]

The British Government was not prepared to tolerate such flagrant violation of the convention which had received the approval and signatures of the German and Italian Governments. It therefore declared that it would employ the Italian hospital ship *Ramb IV*, which had been captured after the fall of Massawa, as a replacement for one of those damaged by the enemy's willful action.[3]

Map 6-1

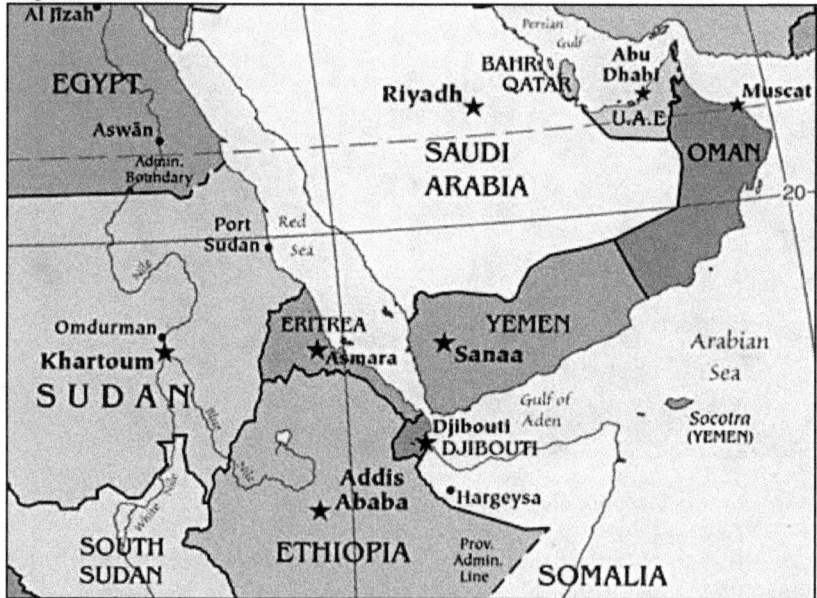

The port city of Massawa lies northeast of Asmara (capital of Eritrea) on the Red Sea

During World War II, the Red Sea Flotilla (part of the Italian Royal Navy) based at the port city of Massawa in the colony of Italian Eritrea, a part of Italian East Africa. Massawa fell on 8 April 1941, after British troops and naval forces attacked it from land and sea. This victory marked the end of the Red Sea Flotilla. Before that day, almost all of its units were destroyed in combat or scuttled by their crews. The auxiliary cruisers *Ramb I* and *Ramb II* and the sloop *Eritrea* departed before Massawa fell, and attempted to cross the Indian Ocean bound for Japan. *Ramb I* was sunk by British warships while the other two ships made it to their destination. As the port city fell, the submarines *Guglielmotti*, *Archimede*, *Ferraris*, and *Perla* escaped. They made the long voyage around Africa and arrived at the German submarine base at Bordeaux, France, on the Bay of Biscay, in May 1941.[4]

The Royal Navy captured the hospital ship *Ramb IV* on 10 April, concurrent with the fall of Massawa. Built in 1938 by the Adriatic Union Shipyards at Monfalcone (located on the Gulf of Trieste in northern Italy), she was the last of four sister ships constructed for the Royal Banana Monopoly to transport refrigerated bananas from Somaliland and Eritrea in Italian East Africa. Their design provided in the event of war, that the ships could be armed and refitted as "auxiliary cruisers" for trade raids. However, when war broke-out, *Ramb IV* was instead converted into a hospital ship for the Royal Italian Navy, for the transportation of wounded Italians to Italy in case Eritrea fell. Such a mission had little chance of success. The Suez Canal was controlled by British forces, and it would have been suicidal for her to attempt to sail around the Cape of Good Hope and to enter the Mediterranean at Gibraltar.[5]

LOSS OF BRITISH HOSPITAL SHIP *RAMB IV*

Photo 6-2

Attacked by German Ju 88 aircraft on 10 May 1942, the British hospital ship *Ramb IV* was set aflame, abandoned, and subsequently sunk by the destroyer HMS *Kipling* (F91). Allison Collection of WWII Photographs

Pressed into hospital ship service by the British, *Ramb IV* served in the eastern Mediterranean. On 10 May 1942, while enroute from Tobruk, Libya, to Alexandria, Egypt, she was bombed by German Ju 88 aircraft, damaged, and set aflame. The violent fire on board resulted in her abandonment; ruined, she was sunk by the destroyer HMS *Kipling* off El Alamein, Egypt. The hospital ship was carrying 360 staff and patients, of which 155 wounded men and 10 crew were lost.[6]

Photo 6-3

German Junkers Ju 88A-5 aircraft captured by the Royal Air Force after it landed in error at Chivenor, on the north coast of Devon, England. It was subsequently flown by No. 1426 Eastern Air Command to demonstrate the features of the Junkers aircraft to Allied aircrews.
Australian War Memorial photograph P00298.003

7

American Forces Land
in French North Africa

In order to forestall an invasion of Africa by Germany and Italy, which, if successful, would constitute a direct threat to America across the comparatively narrow sea from western Africa, a powerful American force equipped with adequate weapons of modern warfare and under American command is today landing on the Mediterranean and Atlantic coasts of the French colonies in Africa.

The landing of this American army is being assisted by the British Navy and Air Force, and it will in the immediate future be reinforced by a considerable number of divisions of the British Army.

This combined Allied force, under American command, in conjunction with the British campaign in Egypt, is designed to prevent an occupation by the Axis armies of any part of northern or western Africa and to deny to the aggressor nations a starting point from which to launch an attack against the Atlantic coast of the Americas.

In addition, it provides an effective second front assistance to our heroic allies in Russia.

—With these words,
President Franklin D. Roosevelt announced the
landing of American troops on African soil on
Sunday, 8 November 1942.[1]

The United States' involvement in the war in Europe began with the Invasion of North Africa on 8 November 1942. Americans first became aware in June 1942 of the planned deployment of military forces overseas. Following a meeting between the Russian Minister of Foreign Affairs and Franklin D. Roosevelt in Washington, D.C., a press release was issued stating that the American president agreed on the "urgent task of creating a second front" that year. Inability by the Allies in Europe to mount an invasion front in Europe had enabled Germany to concentrate its army on the eastern front, leading to a probable conclusion that Russia could not hold out unless something was done quickly to divert German forces elsewhere, such as opening a second operation in Africa.[2]

Germany was already challenging Britain in North Africa. On the heels of this announcement came news of the fall of Tobruk—a port city on Libya's eastern Mediterranean coast, near the border with Egypt—with the advance of German general Erwin Rommel's panzer division into Egypt. Rommel, known as the Desert Fox because of his cunning tactics, was poised to take Alexandria, push the British out of Egypt, and gain control of the Suez Canal. The Allies were thus threatened with both the defeat of Russia and the cutting of the Suez Canal lifeline. The British, in discussions about opening a front in Africa (which preceded the entry of the United States into the war in Europe/North Africa), had envisioned a landing of about 55,000 men in the vicinity of Casablanca, a large port city in western "Vichy" controlled French Morocco on the Atlantic.[3]

Map 7-1

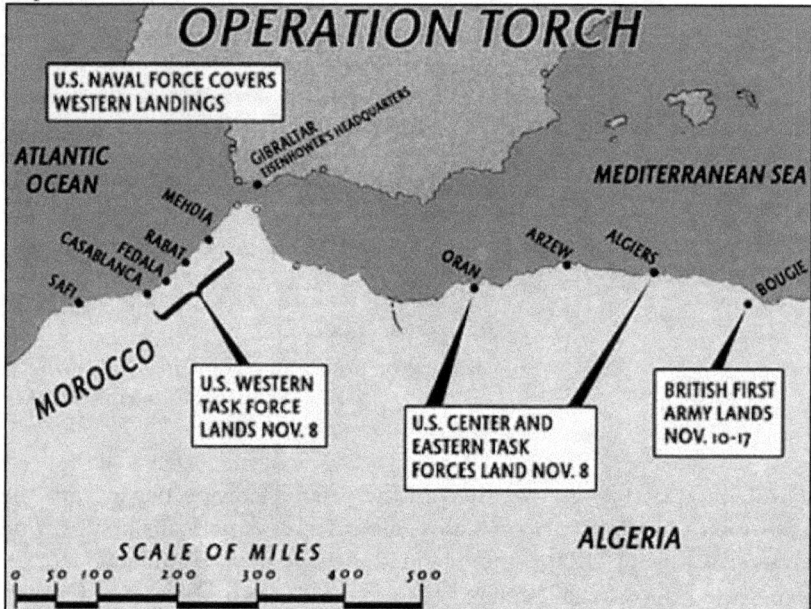

OPERATION TORCH

U.S. NAVAL FORCE COVERS WESTERN LANDINGS

ATLANTIC OCEAN

MEDITERRANEAN SEA

GIBRALTAR EISENHOWER'S HEADQUARTERS

MEHDIA
CASABLANCA RABAT
FEDALA
SAFI

ARZEW ALGIERS
ORAN
BOUGIE

MOROCCO

U.S. WESTERN TASK FORCE LANDS NOV. 8

U.S. CENTER AND EASTERN TASK FORCES LAND NOV. 8

BRITISH FIRST ARMY LANDS NOV. 10-17

ALGERIA

SCALE OF MILES
0   50 100   200   300   400   500

Landing points of assault troops, as part of Operation TORCH, on 8 November, and 10-17 November 1942

After America's entry, the plan was enlarged to include landings not only near Casablanca but also in the Mehdia-Port Lyautey area—a beach village and port on the Sebou River (known today as Kenitra) to the north of Casablanca—and the port city of Safi to the south. Planners then expanded the operation to include occupation of the entire North African coast as far east as Tripolitania, the coastal region of what is today Libya. Occupation by Allied forces of Morocco, Algeria, and Tunisia would help safeguard Mediterranean convoys, thus dramatically shortening the route to the Middle East otherwise around the Cape of Good Hope.[4]

While the United States was to have responsibility for the military and naval operations on the Atlantic coast of Morocco, Oran and Algiers, cities on the Mediterranean on the northern coast of Algeria, were to be captured by two joint British and American forces. For the joint landings, the British were to supply all the naval service except for a few transports while the landing forces were to be partly American and partly British. Allied occupation of all of French North Africa was to be achieved through simultaneous assaults by three attacking forces on Casablanca, and neighbouring locales Oran and Algiers; and shortly after, a separate British one at the port of Bougie, Algeria. Lt. Gen. Dwight D. Eisenhower was given command over the forces, with the exception that British naval units permanently assigned to the Mediterranean would remain under the control of the British Admiralty.[5]

THE MOROCCAN EXPEDITION

The naval component of the Moroccan expedition under the command of Rear Adm. Henry K. Hewitt, U.S. Navy, was designated the Western Naval Task Force. The Army component under Maj. Gen. George S. Patton, U.S. Army, was titled the Western Task Force. The mission assigned to the Naval Task Force was:

> To establish the Western Task Force on beachheads near Mehdia, Fedala and Safi, and support the subsequent coastal military operations in order to capture Casablanca as a base for further military and naval operations.

The objective of the landings at Fedala and Safi was to enable the capture of Casablanca from the land side. Mehdia was to be occupied as a prelude to taking the adjoining airfield at Port Lyautey. At a conference of about 150 naval and army officers convened by Admiral Hewitt at Norfolk, Virginia, on 23 October 1942, the day before the task force sailed, General Patton predicted that all the elaborate landing plans would break down in five minutes, then the Army would take over and win through. He stated in part:

> Never in history has the Navy landed an army at the planned time and place. If you land us anywhere within fifty miles of Fedala and

within one week of D-day, I'll go ahead and win…. We shall attack for sixty days, and then, if we have to, for sixty more.[6]

ATLANTIC CROSSING

Hewitt placed the Western Naval Task Force (Task Force 34) organisation in effect at Norfolk, Virginia, at 0400 on 23 October. Prior to departure, in order to avoid too great a concentration of ships in the Hampton Roads area (where the James, Nansemond and Elizabeth rivers pour into the mouth of the Chesapeake Bay), he moved the Covering Group north to Casco Bay, an inlet of the Gulf of Maine. The Air Group plus three old destroyers being fitted out for special service, were staged at Bermuda. The remaining units, which consisted primarily of combat loaded transports and cargo vessels, with destroyer, cruiser, and battleship escorts, sailed from Hampton Roads in two groups on 23 and 24 October 1942.[7]

Photo 7-1

Adm. Royal Eason Ingersoll, USN, commander in chief, Atlantic Fleet, watches as the Operation Torch task force stands out from Hampton Roads, Virginia, on 24 October 1942, en route to North West Africa.
Naval History and Heritage Command photograph #NH 90944

Once joined, Task Force 34 (also designated Task Force HOW) followed a route to a point south of Newfoundland to give the impression of a regular troop convoy to the United Kingdom. From there to a point south of the Azores, the advance was more or less toward Dakar, the capital of Senegal and an Atlantic port in West Africa. The task force then proceeded toward the Strait of Gibraltar along a track to avoid air search from the Azores and the Canary Islands (should either Portugal or Spain be providing vessel locating information to Germany), until the force split up into three attack groups on 7 November to proceed to final destinations.[8]

Map 7-2

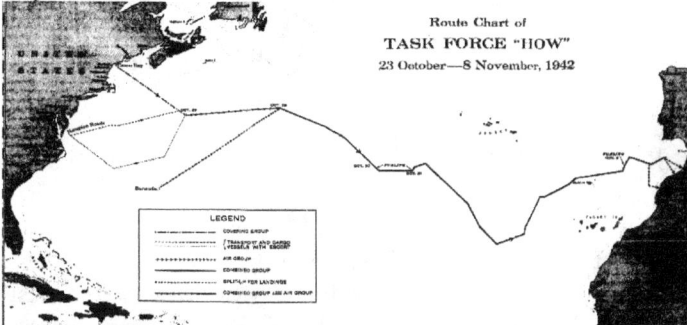

Navigation track of the Western Naval Task Force en route to Operation
TORCH, the Allied invasion of French Morocco
ONI Combat Narratives, The Landings in North Africa, 1944

In execution of Operation TORCH, the Covering Group was to
contain the Vichy French Fleet at Casablanca and, if necessary, the
French units at Dakar, while the attack groups effected landings in the
Casablanca area: at Safi just south; at Fedala, just north; and at Port
Lyautey, further to the north of it. As they neared French Moroccan
waters, the Southern Attack Group left the convoy at daylight on 7
November to proceed to Safi, 125 miles south of Casablanca. In late
afternoon, the Northern Attack Group veered off, headed eastward
toward Port Lyautey, 65 miles north of Casablanca. The Centre Attack
Group continued to Casablanca. As darkness fell, the men aboard the
ships became silent and tense. There was nothing ahead of them, then,
but unknown North West Africa waters and beaches.[9]

Photo 7-2

A U.S. destroyer passes astern of the aircraft carrier USS *Ranger* (CV-4)
at sunset on 8 November 1942, the first day of landings on North Africa.
National Archives photograph 80-G-30232

Details about the successful execution of Operation TORCH, the Allied three-pronged amphibious operation which denied North Africa west of Tunisia to the Axis powers is beyond the scope of this book. The next section is devoted to introducing the entry of U.S. Army Nurse Corps members into the conflict in Europe.

One of the groups of veterans on the front lines in World War II, who have been largely forgotten in the history of that great conflict, were the women who served and died doing their duty alongside the men of the United States Army. These were the women of the U.S. Army Nurse Corps, who served from the first day of the war to the last, suffering deaths and wounds as they treated the soldiers, sailors, airmen, and civilians who were wounded or sick.[10]

Photo 7-3

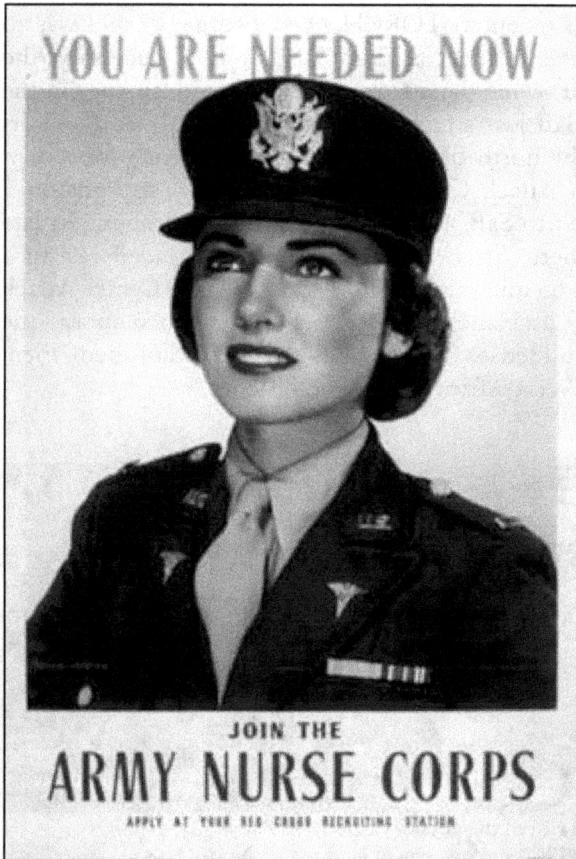

World War II U.S. Army Nurse Corps recruiting poster.

U.S. ARMY NURSES LAND IN ALGERIA

The 48th Surgical and 128th Evacuation Hospitals are one and the same unit. The 48th Surgical Hospital, a 400-bed medical unit arrived in the MTO Theatre [Mediterranean Theatre of Operations] on 8 Nov 42 and served in Algeria and Tunisia. Campaign credits include Algeria, French-Morocco, and Tunisia. On 1 May 1943, the unit was reorganized and re-designated the 128th Evacuation Hospital, Semimobile, still a 400-bed Hospital which continued to serve in Tunisia and Sicily, until it got transferred to the ETO [European Theatre of Operations] on 26 November 1943. Additional Campaign credits were obtained including Sicily, Normandy, Northern France, Rhineland, Ardennes-Alsace, and Central Europe.

—48th Surgical / 128th Evacuation Hospital Unit History.[10]

On 6 August 1942, the Nurses of the 48th Surgical Hospital left the United States bound for the UK aboard USS *Wakefield* (AP-21), sailing from New York as part of Convoy AT-18, the largest troop convoy at that point yet assembled. A dozen troop transports made up the bulk of the convoy, escorted by twelve cruisers and destroyers. The hospital personnel disembarked at Greenock, Scotland, on 18 August; then traveled to southern England, reaching London on 20 August. At this point the unit was made up of the following numbers:

- 22 Medical Officers
- 1 Chaplain
- 60 Nurses
- 250 Enlisted personnel[11]

Photo 7-4

USS *Wakefield* (AP-21) under way, 11 May 1942, near the Philadelphia Navy Yard.
National Archives photograph No. 19-N-29517

The unit was initially based in Tidworth (in southeast Wiltshire, England, on the eastern edge of Salisbury Plain) for almost 2 ½ months, with days spent marching, drilling and guarding. It then moved north into the mountains outside of Glasgow to begin a whole new regimen of field living and training.[12]

On 23 October 1942, when Convoy KMF-1 sailed from the UK bound for North Africa, the 48th Surgical Hospital was embarked aboard HMS *Orbita*, HMS *Monarch of Bermuda*, and USAT *Santa Paula*. Its total complement was 48 officers, 57 nurses, and 273 enlisted men. In early morning on 8 November (D-Day), the nurses and the other members of the unit climbed over the side of a ship off the coast of North Africa and down an iron ladder into small assault boats. Hospital personnel were equally distributed with each boat carrying 5 nurses, 3 medical officers, and 20 enlisted men.[13]

The nurses wore helmets and carried packs containing musette bags, gas masks, and canteen belts. Only their Red Cross arm bands and lack of weapons distinguished them from fighting troops. They waded ashore near the coastal town of Arzew with the rest of the assault troops and huddled behind a sand dune while enemy snipers shot at anything that moved. That evening they found shelter in some abandoned beach houses, a somewhat safe haven in which to rest.[14]

Before the night was over, however, their commanding officer ordered them to an abandoned civilian hospital, where they began caring for invasion casualties. There was no electricity or running water, and the only medical supplies available were those the nurses had brought themselves. Enemy air attacks on the harbour at Arzew would delay the unloading of supplies for two days.[15]

The hospital came under sporadic sniper fire as doctors operated by light from flashlights held by nurses and men of the medical unit. As there were not enough beds for all the casualties, wounded soldiers lay on the concrete floor in pools of blood. Nurses dispensed what comfort they could, the only sedatives available being what they had carried with them during the landing.[16]

LESSONS LEARNED / PRACTICES INACTED

U.S. Army nurses and other medical staff serving at the front in North Africa became expert at meeting the challenges of combat while caring for incoming patients. In February 1943, when news reached the 77th Evacuation Hospital bivouacked near Tebessa in northeastern Algeria, that the German Army had broken through the Kasserine Pass in west central Tunisia, staff members packed up and moved their 150 patients sixty miles to a safer bivouac. Within twelve hours a new hospital was fully operational and had received 500 casualties.[17]

The Army Medical Department's newly organised "chain of evacuation," and the nurses' role in it, were tested in North Africa and ultimately used successfully in every theatre in the war. Mobile field and evacuation hospitals, which closely followed the combat troops, were usually set up in tents and were subject to move at short notice. Nurses packed and unpacked these hospitals each time they moved.[18]

Usually, 18 nurses were assigned to field hospitals, which could handle 75 to 150 wounded patients brought to them by litter bearers and ambulance drivers. Doctors and nurses performed triage on arriving patients at the receiving tent to ascertain the severity of their condition. A need for special treatment determined when, how, and where a patient was to be sent. Those judged strong enough to travel were taken by ambulance to evacuation hospitals farther away from the front lines and near transportation facilities.[19]

Photo 7-5

Drawing "Chart Accompanies the Patient" by Joseph Hirsch, circa 1943.
Naval History and Heritage Command photograph

Nurses stabilized patients requiring treatment elsewhere with blood, plasma, medication, and dressings before sending them on. Those who needed immediate care went directly into surgery. Others who needed surgery but were too weak for an immediate operation and could not travel, were sent to the shock ward.[20]

Photo 7-6

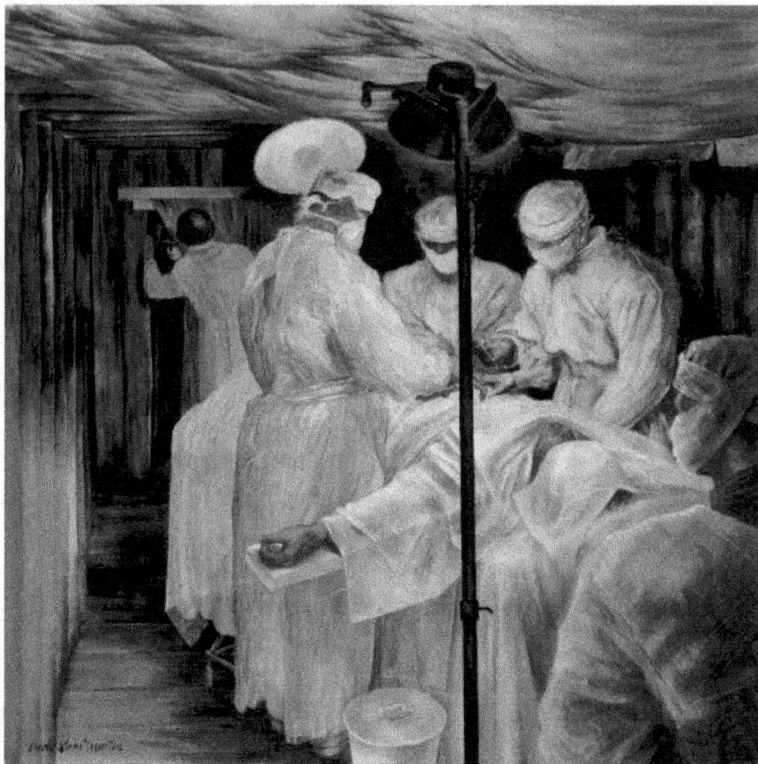

Painting "Life in Death" by David Stone Martin, circa 1943.
Naval History and Heritage Command photograph

Typically, a field hospital could perform approximately eighty operations a day, with over an 85-percent survival rate. When postoperative patients were strong enough for travel, they were transported by ambulance to evacuation hospitals.[21]

Evacuation hospitals, assigned 53 nurses, could accommodate up to 750 patients. Their doctors operated as necessary on patients sent from field hospitals. Those with postoperative stomach wounds were routinely kept in an evacuation hospital ten days before they were sent on, and those with chest wounds usually kept at least five days.[22]

Where facilities were available, critically wounded patients needing specialized treatment were evacuated by air to station and general hospitals. Other patients were evacuated from the field to station and general hospitals via hospital trains, hospital ships, and aircraft attended by nurses and other medical personnel. On hospital trains, nurses dispensed medication and food to their patients, made them as comfortable as possible, and monitored them for signs of stress and complications. Such trains usually had thirty-two beds per car, with one nurse devoted to each car of litter patients or to several cars of ambulatory patients.[23]

Photo 7-7

Drawing "Transport Wounded" by Kerr Eby, 1944.
Naval History and Heritage Command

Photo 7-8

Ambulance Train, St. Nazaire, France.
Naval History and Heritage Command photograph #NH 115700

Station hospitals received battle casualties from evacuation hospitals and performed surgery and specialized treatments; whereas general hospitals were the last step in the evacuation line. Patients still requiring diagnosis, specialized lab tests, or long periods of recuperation and therapy were sent to general hospitals. Upon release, patients were either returned to duty or sent back to the United States.[24]

Photo 7-9

Drawing "Hospital Ward" by Carlos Andreson, 1943.
Naval History and Heritage Command

Photo 7-10

Painting "U.S.O. Show" by Carlos Andreson, depicting at least two attendees with crutches, 1943.
Naval History and Heritage Command photograph

8

First U.S. Army Hospital Ship Arrives in European Theatre

Photo 8-1

Portrait of Lt. Gen. Dwight D. Eisenhower, U.S. Army, by McClelland Barclay, USNR, circa 1943. (Eisenhower received his fourth star on 11 February 1943.) Naval History and Heritage Command photograph #NH 78357-KN

When the United States entered World War II, following the Japanese attack on Pearl Harbor on 7 December 1941, the Navy had only two hospital ships, and the Army none whatsoever. This glaring deficiency was highlighted by the inevitable personnel casualties resulting from the North African and Italian campaigns of 1942-1943.[1]

Pending the procurement of regular hospital ships, most Army patients in 1942 were evacuated from overseas areas by means of returning Army vessels. The emphasis at this time was on getting troops overseas rather than on bringing back casualties. Other means of evacuation were also considered in 1942. These included the possible use of ambulance transport, the provision of additional hospital space on designated British vessels, and evacuation by air.[2]

In 1942, U.S. Navy hospital ships USS *Relief* (AH-1) and USS *Solace* (AH-5) were occupied with operations in the Pacific. Accordingly, for pending operations in North Africa, the Army had to rely upon its own resources.[3]

On 6 August 1942, Lt. Gen. Dwight D. Eisenhower was designated Supreme Commander, Allied Expeditionary Force. During the planning of the North African operations, Eisenhower repeatedly called attention to the need for Army hospital ships. On 21 August 1942, he reported to the Adjutant General at Washington, D.C. that three 500-bed hospital ships would be required by late September and that modified passenger vessels would be acceptable. The acute shortage of transport ships, however, precluded the conversion of any for service as Army hospital ships in the early stages of the North African campaign.[4]

Subsequently, after the landings by United States forces in North Africa, it was found possible to make alternative arrangements by utilising British hospital ships for evacuation of U.S. casualties to the United Kingdom. Nevertheless, on 27 March 1943, the commanding general, North African Theatre (Eisenhower), notified the War Department of increased requirements that could not be met by these arrangements for sea evacuation of patients, and requested two United States hospital ships.[5]

THE U.S. ARMY'S FIRST HOSPITAL SHIP IN WWII

Photo 8-2

U.S. Army hospital ship *Acadia* under way; location and date unknown.
Harold Larson, *Army Hospital Ships in World War II* (Washington, D.C.: Office of the Chief of Transportation, Army Service Forces, 1944)

The pressing need of top Army combat leaders for hospital ships was shared by its medical establishment. In connection with extensive combat operations in North Africa and in the Mediterranean, the Surgeon General of the United States requested hospital ships for returning sick and wounded veterans to America. The *Acadia*, a 402-

foot vessel built in 1932, and initially a troop ship, became the means adopted for this service. It was the first U.S. World War II vessel so commissioned under rules of The Hague Convention applicable to hospital ships.[6]

During World War II, *Acadia* initially after being employed as a troopship, served as an ambulance transport, before her conversion to a hospital ship. Her Army history began on 25 November 1941, when the Service obtained the 6,185-ton vessel built by Newport News Shipbuilding and Drydock Co., Newport News, Virginia, on a time charter. Following hasty conversion into a transport at New York, she left on 20 December bound for Balboa, Canal Zone. She subsequently operated in the Caribbean area before ordered to Boston for conversion to ambulance transport.[7]

The objective of the conversion to "ambulance transport" was not to produce a true hospital ship, but instead a hybrid vessel. The plan was to sacrifice minimum troop-carrying capacity on outbound trips in order to provide an approximate 500-patient capacity for bringing combat casualties home on return trips from Europe and North Africa. The conversion was begun on 11 June and was completed on 16 October 1942.[8]

Acadia arrived at New York the following day, 17 October 1942, and was inspected by the Port Surgeon, Col. Louis A. Milne. He identified that the quarters available for both the nurses and enlisted personnel were inadequate. Necessary work to correct this deficiency was performed at New York, and the ambulance transport was ready on 4 December 1942.[9]

Acadia made several voyages between New York and North Africa, between December 1942 and April 1943, sailing in convoy and carrying troops outbound and military casualties inbound. In these duties she maintained the garb of a troop ship. Early in 1942, although the U.S. Army had no hospital ships as yet, the Surgeon General's Office drafted a Table of Organisation for a Medical Hospital Ship Company. It stipulated that Army hospital ships were to be manned and operated by civilian Merchant Marine personnel and by Army Medical Department personnel to care for patients.[10]

Acadia departed on her first voyage with a full hospital complement aboard, the 204th Medical Hospital Ship Company consisting of 18 officers, 37 nurses, and 94 enlisted men. She travelled in convoy unmarked as a hospital ship. The Atlantic crossing was the 204th's introduction to sea duty. When the ship pitched and rolled, and seasickness began to take its toll, many a meal was lost by soldiers and nurses. Arriving at Casablanca, French Morocco, there was only time for those on board to get their first glimpse of a war-torn country as patients were loaded; and then it was back to New York. (Summary information about numbers of patients evacuated from overseas and debarked at U.S. Army Ports may be found in Appendix D.)[11]

For the next four months, *Acadia* carried the hospital unit many times again to North Africa with its blazing sun, odd foul smells, and its strangely dressed natives. The ship crossed between New York and North Africa, carrying troops on the outbound trip and wounded patients on the return voyage. In March, the urgent need of hospital ships in the North African Theatre led to a demand for a change in *Acadia*'s status/capabilities from ambulance transport to full hospital ship.[12]

CONVERSION TO HOSPITAL SHIP

All of the hospitals in the area were full and overflowing with patients and wounded. [Furthermore], there was dire need for evacuation of a great many of these men to relieve the congestion of hospitals and to make more room for the wounded coming in hourly. [Yet it was decided at Oran that only ambulatory patients could be removed on the Acadia *to the United States. As a result] this excellently equipped ship had to return practically empty [despite the presence of trained personnel (the 204th Hospital Ship Company) to care for a full load of the sick and the wounded.]*

—Report by the Surgeon aboard *Acadia*, dated 11 March 1943.[13]

The chief surgeon aboard *Acadia* had highlighted in a report that the ship was not being fully utilised, because patient transport to the United States was much below her capacity due to a decision that only ambulatory patients could be carried. Further, he noted that the hospitals in North Africa were in great need of supplies which could be brought on board *Acadia* without violation of any treaties, and lamented that his personnel were "traveling needlessly back and forth on what might be called a pleasure cruise."[14]

By April 1943, it became evident that *Acadia* could be operated more safely and expeditiously for evacuation of the sick and wounded if registered as a regular hospital ship, rather than continuing service as an ambulance transport. The decision to employ her as a regular hospital ship resulted in a secondary conversion accomplished at New York, from April to June 1943.[15]

Acadia's hospital ship transformation included the removal of her antiaircraft and other guns (replaced by some small arms: a few automatic pistols, 5 shotguns, and 3 outmoded bolt-operated rifles); the Navy crew and troopship bunks; and the exchange of her gray war paint coat for a white and green one prescribed under international practice to identify hospital ships. The latter involved painting the entire hull white, with a horizontal green band the whole length of each side, and the usual red crosses on the sides, deck and funnel, which could be illuminated at night. After being duly registered under the Treaties of The Hague Convention, the new United States Army Hospital Ship *Acadia* was ready to sail once more.[16]

MAIDEN VOYAGE OF USAHS *ACADIA*
On 5 June 1943, U.S. Army hospital ship *Acadia* left New York on her first voyage to Oran, Algeria. Lt. Col. Thomas B. Protzman, MC, (commanding officer of the 204th Medical Hospital Ship Company) described her departure:

> We were supposed to sail at 0700, but there was a large convoy going out so we had to wait until they had cleared the lightship. At 1330 we cleared the dock and started our maiden trip as a Hospital Ship. Two Army blimps followed us until dark to ensure our safe passage through the submarine area.[17]

Comprising the hospital ship's U.S. Merchant Marine and U.S. Army Hospital Company leadership were:
- Captain "Jack" John W. Kirchner (Master)
- James L. Kane (Chief Engineer)
- Lt. Col. Thomas B. Protzman, MC
  (Commanding Officer, 204th Medical Hosp Ship Company)
- Maj. William V. Barney, ChC (Chaplain)
- Maj. William G. Ford, MC (Chief Medical Staff)
- Capt. Louis V. Angioletti, MC (Chief Surgical Staff)
- Capt. Alexander S. Forster, DC (Chief Dental Staff)
- 1st Lt. Muriel M. Westover, ANC (Chief Nurse)[18]

Everyone on board including the civilian Merchant Marine crew had to be certified neutral (for protection by The Hague Convention), and carry a certificate to that effect with a photograph and fingerprints of each individual.[19]

On 8 June, Captain Kirchner issued the then-new self-inflating life vests to all hospital unit personnel and to all crew members. Much smaller than the kapok life jacket, the vests were rubber reinforced with cloth and wrapped around the waist.[20] (The author would rather have a kapok, as they don't fail if unable to inflate or hold air.)

Nearing the Straits of Gibraltar in early morning on Sunday, 13 June, the hospital ship encountered a German U-boat on the surface, which left her unharmed. Protzman described the incident, and *Acadia*'s ensuing short call at Gibraltar:

> I got up at 0430 this morning so as not to miss the entrance to the Straits of Gibraltar and shortly after I reached the bridge, a German submarine came up on the starboard bow and followed us for nearly a half hour but did not attempt to either hail or stop us. At about 0830 we entered the Straits and the mist was so heavy that we could scarcely see either shore, but this cleared by 0900 as we now came close to the Spanish coastline.

> After religious services, we all went on deck and had our first clear view of The Rock. As we passed the fortress and were heading into the Mediterranean, the British sent us a blinker message to head into the Bay and be identified. A pilot was sent aboard and we were safely guided through the minefields and anchored in Spanish waters just off the little town of La Linea, at the base of The Rock and right alongside the British Hospital Ship HMHS *Oxfordshire* (which I visited before at Mers-el-Kebir on my last voyage).

> I had time for a short trip into The Rock and visited the town near the airport. All the time that we were anchored Spaniards came out in small fishing boats and tried to sell some vile liquor to the sailors. The Master drove them off with the fire hose. At 1700 the *Acadia* was on her way and we hope to reach Oran, Algeria, by 0900 tomorrow.[21]

Photo 8-3

Gibraltar from the northwest.
Naval History and Heritage Command #NH 1444

Map 8-1

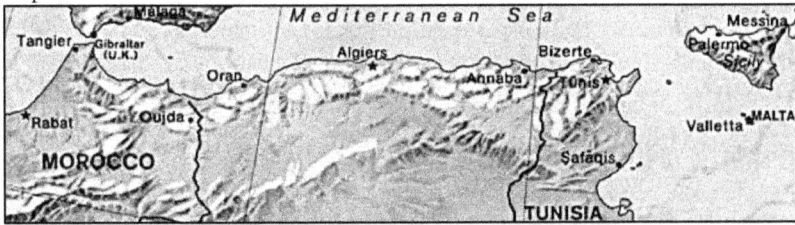

Southeast area of the Mediterranean and areas of bordering countries;
Mers-el-Kebir (not shown) is located about 14 miles west of the coastal city of Oran.

Around 0700 the following morning, 14 June, the cliffs of Oran, Algeria, came into view with the haze-covered mountains in the background. *Acadia* had been ordered into Mers-el-Kebir and was to berth at the tip of the jetty about fourteen miles from Oran. HMHS *Oxfordshire* has been ordered in also, providing Protzman an opportunity to renew old acquaintances. He witnessed while entering port, evidence of forthcoming combat operations:

> As we came into Mers-el-Kebir a long line of assault landing barges was on its way out towards Algiers; something is getting very hot in the area as everything is being moved up into Tunisia. In spite of the fact that thousands of wounded have been shipped out of this area, there are still almost 11,000 patients in the Oran area alone, and a considerable number near the front. We saw H.M. King George VI in Oran today. Our orders are to move to Oran and load at once, for a [planned] quick return trip to Charleston [after first arriving at New York], and right back again. This looks more than ever as though we will get back in time for the invasion [of Sicily].[22]

Photo 8-4

Mers-el-Kebir, Algeria, looking across the fortress and harbour, with the city of Oran in the left distance, circa 1943-1944.
National Archives photograph #80-G-K-637

The following evening, 15 June, Protzman and other Hospital Unit personnel were invited to a GI banquet (Army social event) with French wine. The Nurses wore evening gowns sewed from cloth they were able to buy from the natives. Everyone had a splendid time and drove back early in the morning with lights blacked out (extinguished) ashore. When they reached the *Acadia*, nearly 500 patients had been loaded and the rest were expected later that morning.[23]

Acadia left the pier at 1700 that evening, 16 June, to begin her independent return voyage to the United States. Protzman described the ten-day transit in his diary, some excerpts follow:

16 June: The sea is pretty rough and the wind is blowing so hard I can hardly hold my camera steady. We should pass The Rock of Gibraltar about 0600 in the morning and be well out in the Atlantic by the afternoon. We travel alone and our boat will be routed the shortest way home.

17 June: The sea is pretty rough and most of the ambulatory cases are getting seasick. Some of the boys have started to gamble openly in rather large games so I issued a stop order on the playing and appointed one of my staff as Provost Marshal to carry out the job.

18 June: We have been doing approximately 19 knots all day even though the weather continues to be rather rough. The day was uneventful until the evening when during a jam session 3 of the NP [neuropsychiatric] cases began to fight and we had lots of trouble getting them under control and into restraining sheets. This was only possible after they had been given large doses of Sodium Amytol and Paraldehyde. One of the merchant mariners had sold them some liquor and that is what started the fight. We have 263 psycho cases on board and no adequate place to hold them (there are only two padded cells in the stern of the ship). One patient Officer admitted paying as much as $100.00 for a quart of rye whiskey, but couldn't put the finger on the sailor when I had them all paraded for investigation.

19 June 1943: The Azores are only about 60 miles away, so maybe the boat will reach there. Everything is going fine on board; the ARC [American Red Cross] workers and the Physiotherapists made the boys more comfortable and kept them occupied. One poor guy in my surgical ward had both his eyes shot away and the Red Cross girls have been wonderful in their efforts to bring up his spirits and to give him some hope for the future. In another ward, I discovered a boy who had been an accordionist in a famous dance orchestra stateside. He has severe shrapnel wounds in both hands sustained

at Bizerte, Tunisia. Only he and his Captain came out alive. He was sitting alone in a corner of the ward crying; the only thing he knew was music, his only means of support, and his fingers were all stiff and sore. One of the Physiotherapists has baked and massaged his fingers and hands and today, the patient was able to play a tune on the ship's accordion.

20 June: Of the 800 patients we have on board, only 150 have service records. Our Chaplain did a good job this morning; his sermons are short and snappy but have a lot of meat in them.

21 June: The little merchant mariner (manic depressive) has just finished ripping his sixth bathrobe into shreds. None of the sedatives seem to help control him. The NP case who started the fight is still in restraining sheets and will probably remain so for the rest of the voyage. The whiskey certainly did a tremendous amount of damage to his mental system; he has been raving wild ever since and threatens to kill us all if he ever gets out. There are 2 men on guard at his bed all the time. At about 1730 we sighted a large iceberg, several times the size of this ship. The barometer has been dropping all evening, but at 2130 the Captain called me and said to secure all patients as we were running into a hurricane. It effectively hit us about 2200 and the ship really did some stunts. We just kept going headway into the storm at about 3 knots; none of us got much sleep, but we managed to get through all right without any serious damage. We will probably be a day late for arrival.

22 June: The sea is considerably calmer and we are making about 6 knots but the percentage of seasickness has risen to appalling proportions. As we depend on ambulatory patients to help with cleaning, dishwashing and guard duty, the ship will just have to get and stay dirty and go without guards. If nothing special happens we should be in by next Friday and I certainly hope we aren't kept out in the harbour as last time.

23 June: We are pretty well out of the storm area and there is much less sickness, but we have lost a whole day because of it and won't arrive until Friday morning.

24 June: Clear calm day with a bright sun. A great many of our patients have been on deck long before daylight just in the excitement of being only a few hundred miles from home. Even our mental patients are better, yet the poor devils won't get the freedom we give them here. We can afford to be more generous at sea as they have no place to escape to. The boy with the accordion is playing quite well now and his face just shines with hope. The little

merchant mariner's situation improved and we allowed him out of his dell for the first time yesterday.

25 June: New York, N.Y., Port of Embarkation, Zone of Interior. We didn't reach "Ambrose Light" [Ambrose Lightship WAL 533, navigational light beacon marking Ambrose Channel, one of the main shipping channels to New York Harbor] until 1030 and will not dock until well in the afternoon. The skipper called the Port Authority and we are to dock at Pier 15 for debarkation.[23]

Upon the hospital ship's arrival at the pier, an inspection party headed by the Surgeon General of the Army, the Chief of Transportation, and the Commanding General of the New York Port of Embarkation boarded the vessel. In a letter dated 26 June 1943 the Chief of Transportation advised the Commanding General, New York Port of Embarkation, that he and the Surgeon General both felt "that there was something wanting on the *Acadia*." The most grievous deficiencies were (1) that the nurses wore regular working slacks that varied in colour and quality, and (2) many of the patients were badly in need of haircuts and shaves.[24]

Lieutenant Colonel Protzman recalled about the inspection:

Major General Norman T. Kirk [The Surgeon General], with his entire staff was there. He made a quick tour of the ship and seemed satisfied. Major General Charles P. Gross [Chief of Transportation, Army Service Forces], was quite critical of my Nurses because I had them dressed in slacks. He also thought my EM [enlisted men] were not dressed properly; but when I advised him that he was looking at my NP [neuropsychiatric] patients instead of at my soldiers he withdrew his statement. I also told him that we had just come through a two-day hurricane and were more concerned with the care of our patients than with the preparation of a parade.[25]

(In follow-on corrective action, the Office of the Surgeon General arranged to have suitable standard slacks manufactured and issued so that the nurses would present a uniform appearance. For the patients, a barber shop was established aboard the ship to provide needed haircuts and shaves.)[26]

Acadia had carried 749 Army and 29 Navy patients on her return voyage from Oran. Following her arrival at New York on 25 June, the Army patients—195 medical, 295 surgical, and 259 neuropsychiatric cases—were evacuated and delivered to the Halloran General Hospital on Staten Island.[27]

Photo 8-5

Halloran General Hospital postcard.

Protzman learned following the arrival of *Acadia* at New York on Friday, 25 June, that the ship was going to make a quick turn around and leave again on Sunday or Monday. He subsequently noted about this change in orders:

> Something very hot must be cooking and we must be badly needed. The USAHS *Seminole* the SECOND Army Hospital Ship, sailed for North Africa on Monday of this week. When I inspected her the day we sailed [on *Acadia*'s previous voyage], the whole ship had been gutted, and at the time I felt that she couldn't possibly be finished for at least a couple of months. Yet, Colonel Holder [Office of The Surgeon General] had produced another miracle and we will now have 2 Hospital Ships in the combat zone. If the new CO runs into the storm we just came through, he will get a baptism of a very unpleasant character. Its capacity is about 450 wounded.
>
> While the Surgeon General was visiting, we clarified the lifesaving situation and will do as the British – carry our full capacity of absolute litter cases. This means of course that in the event of a torpedo or mine, or another disaster, the badly wounded cases will have to take that extra chance. They are given the protection of the Geneva Convention and that will have to suffice as it is imperative that those totally disabled with wounds and injuries be removed from the Theatre of Operations as soon as possible.[28]

Lieutenant Colonel Protzman spent the entire day on 27 June, Sunday, with "the powers that be, of the Port and Washington." *Acadia* now had definite orders to sail Monday night without fail. As a result of new policy, he was to be in command of the hospital ship (as well as the hospital on board her) for the forthcoming operations. Protzman recorded in his diary, his hopes and fears regarding the impending voyage, and some final preparations before departure from New York:

We are to take part in the attack by going in with the Allied invasion forces. Time and place are a deep secret. This next trip, I hope, will make favourable history for the Medical Corps, as a new policy has been instituted; indeed, this is the first time in the history of the United States, that an MC Officer has been given complete command of the ship and the Hospital. Before, a line Officer was given command of the ship, and a Medical Officer command of the Hospital unit, subservient to the line Officer.

I pray the Gods that luck be with me, for on my actions depends the future policy of the War Department concerning the powers of the Medical Department in all future ventures of this magnitude. Though we have never been in actual battle, I feel and am confident that my outfit will come through as they have in previous emergencies.

I had to keep all Officers, Nurses, and Enlisted personnel on the boat all day as we weren't sure just when the sailing time would be.... Practically all the Officers, the Captain and myself went to "Gloucester House" on 51st Street and had lobster for supper. We then returned to the boat and began to check on loading the extra supplies.[29]

On the morning of Monday, 28 June, USAHS *Acadia* left the pier in New York at 0535, bound for the Mediterranean on her second voyage as a hospital ship. Protzman noted about the uncertain future, "We only have food enough for three months so there's a possibility that we may return by then."[30]

USAHS *Acadia*
A White Ship
(by Walter D. Higgens, Captain, MC, U.S. Army, 24 June 1943)

She Lies at Anchor in the Bay
A Big White Ship looking Gay
Amidst her Sister Ships of Gray

A Big Red Cross upon her Sides
Marks her Mission as She Rides
Upon the Water and the Tides

Within her Decks are Those in White
Who Heal our Wounds Throughout the Night
And Give Those Strength Who Fought the Fight[31]

9

USAHS *Acadia* and *Seminole*'s Support of Operation HUSKY

USAHS *Acadia* left New York on 28 June 1943 on her second voyage, bound for the Mediterranean. The next morning, Lt. Col. Thomas B. Protzman MC, (commanding officer of the 204th Medical Hospital Ship Company) held a meeting with the officers and nurses, concerning his planned utilisation of personnel, and administration of medical care:

> Operating teams were formed, of which there will be three; shock teams of Enlisted Men and Nurses for plasma and blood transfusions were established. The EM [enlisted men] will handle the general nursing while the Nurses will be stepped up to do the immediate first aid work, thus permitting the doctors to act on the surgical or burn teams. Our Chaplain will take over supervision of loading of the wounded by means of metal "Stokes" litters. The latter task may be very difficult as the operation may have to be accomplished from the water or from lighters or landing craft and if there is any sea at all the job will be hazardous! The ARC [American Red Cross] workers, Physiotherapists and Dietitian will attend to the feeding of hot drinks and sandwiches to those cases able to take nourishment and to assist the shock teams. For 8 hours each day, everyone will be making and preparing surgical dressings and supplies.[1]

On 3 July, although following a southern route to escape fog and icebergs, *Acadia* encountered heavy fog and with her whistle blowing at regular intervals, she plugged along at only 4 knots; then ran into a fairly large area of slush ice. She had orders to be in Gibraltar by 7 July, for receipt of further instructions. So, once clear of navigational hazards, *Acadia* made best possible speed to make up for the lost time.[2]

At noon on the 4th—America's Independence Day—*Acadia* was only 800 miles from Gibraltar. Then, being ahead of schedule, she slowed down. On 5 July, a four-engine B-24 Liberator bomber on patrol overflew the ship, then returned in the direction of Gibraltar. Unless fuel conservation is a concern, ships typically like to be a little ahead of schedule, in the event of unforeseen difficulties. Because she did not want to enter the Straits of Gibraltar before daylight, the early morning of the 7th found *Acadia* making circles, marking time off the entrance to the Mediterranean.[3]

She entered the Straits around 0700, and nearing The Rock as the prevailing fog cleared, it was possible for those aboard to see several aircraft in the sky and a harbour full of ships. *Acadia* anchored at 1100 near the USAHS *Seminole* and three British hospital ships. As noted in a diary entry, Protzman believed that combat operations were near at hand, but the ship's only firm orders then were to proceed to Algiers, Algeria:

> It looks like the real thing now, and all preparations won't be in vain. The Port Commander left us at noon after giving us instructions to move up to Algiers the day after tomorrow. On talking with some of the British Officers, I find that the plan is to sandwich us between two British ships so that we can have the benefit of their previous battle experience. From the looks of things, I would say the invasion will start within the next few days.[4]

At 2000 on the night of their arrival, Protzman had some of his supplies delivered to *Seminole*, the U.S. Army's second hospital ship in World War II. Protzman (who, neither a mariner nor a naval officer, routinely referred to ships as boats) had empathy for *Seminole*, and the challenges she faced having been rushed into service:

> These poor devils are in a very bad situation. Their vessel was fitted out so rapidly [May-June 1943] that they have no sufficient supplies and a great deal of their equipment hasn't been installed. When their boat first sailed from New York, they were to go to Oran, Algeria, and pick up a load of wounded and come right back. Now they have been ordered to an invasion point and they aren't ready, through no fault of their own.[5]

At 0200 on 8 July, Protzman was awakened by the sounds of several depth charges going off. Groggy, he thought at first that he was aboard a ship in convoy, with escort vessels dropping explosives on enemy submarine contacts. Once fully awake, he found that it was only a British patrol boat dropping charges in the bay to discourage any saboteurs or infiltrators. The plan to sandwich the US ships between British ones must have been abandoned as the British hospital ships departed Gibraltar in predawn darkness before the new day broke.[6]

Seminole was scheduled to leave midafternoon for Bone, Algeria. Before she sailed, Protzman sent her commanding officer Lt. Colonel C. W. Salley and Chief Nurse Catherine Ambry additional supplies, and noted about Salley's other difficulties:

> He hasn't been able to get his motor boat fixed. All his life boats are hand-operated and neither the merchant marine crew nor his medical complement know how to operate them. I certainly hope he won't get into serious action.[7]

At the time, although the U.S. Army had recently designated medical officers in charge of medical units aboard hospital ships, the commanding officer of the ship as well as its medical personnel, the Merchant Marine master who operated the ship, and for whom the crew worked, was still rightly addressed as Captain.   A diary entry by Protzman evidences this, and notes his displeasure with the fare ashore that night:

> The Captain and I were trying to get more [fuel] oil for our ship and we spent a lot of time trying to unroll British red tape; we finally got the oil after seeing the American Consul. We stopped to have dinner at the "Hotel Bristol" and paid $2.00 per person for some lousy food and warm beer.[8]

At noon the following day, 9 July, *Acadia* weighted anchor and, amid excellent weather, stood out of Gibraltar, and proceeded eastward in the Mediterranean.[9]

INVASION OF SICILY – OPERATION HUSKY

We started the day [10 July] by listening to the news reports of the assault against Sicily and all hope our boys can hold their beachhead.... 1030, we just passed a large convoy of Liberty ships going into the Mediterranean loaded with troops and supplies.... At 1500 hours we were anchored well in Algiers harbour away from the city, and the sea is too rough to attempt to go ashore by small boat. While at anchor a fully-loaded convoy left for the east, crossing the one we passed earlier in the day. There were thirty-seven ships in this one. The harbour is full of troop and cargo ships fully loaded and ready to move where needed. We heard there were 1,200 ships during the assault against Sicily. 1600 hours and another convoy comes in. Because of the ongoing sea traffic, we will remain offshore and out in the ocean till tomorrow. The harbour is still protected by barrage balloons. The last enemy air raid took place on 15 June and I believe the Krauts have too much other business to attend than to bother us tonight, though the moon is very bright.

—Lt. Colonel Thomas B. Protzman describing *Acadia*'s second day out
of Gibraltar, and arrival that day (which marked the commencement
of the Allied invasion of Sicily, Italy) at Algiers, Algeria.[10]

Following the surrender of the Axis forces in North Africa, the
Americans were eager to make a cross-Channel invasion, with the
purpose of drawing the Germans into a decisive battle that would end
the war. The British, however, felt that the combined Allied forces were
not yet ready to face the Germans in France, arguing instead for an
assault on the Continent out of the Mediterranean into what Churchill
called the "soft underbelly." The British argued that the Germans were
weaker there and attacking this soft underbelly in either Italy or the
Balkans would relieve pressure on Britain as well as give the Americans
more combat experience before they confronted the main body of the
German Army. It would also draw German divisions from the Eastern
Front, which would serve to fulfill Soviet Union leader Joseph Stalin's
demand for an Allied second front in Europe.[11]

Another argument that favored an invasion from the
Mediterranean was that Italian resolve was wavering and if they could
be taken out of the war it would ease Allied efforts considerably. There
was also an active Italian partisan movement that would harass the
German rear echelons. The decision was made to invade Sicily since
seizure of the island would provide a steppingstone to the Italian
mainland.[12]

There were about 250,000 Axis troops defending Sicily, most of
them Italians with a stiffening of German troops and a large number of
Luftwaffe personnel. The coast defence formations were intended to
act as a delaying force, giving time for the regular Italian divisions and
the Germans to react decisively. The largest German units present for
the landings were the Hermann Göring Panzer Division and the 15th
Panzergrenadier Division. Between them they had about 150 tanks,
including 17 heavy Tiger tanks together with infantry and supporting
artillery.[13]

On the night of 9-10 July 1943, Allied forces commenced
Operation HUSKY—the code word for the invasion of Sicily—which
launched the Italian Campaign. This large scale amphibious and
airborne operation, with simultaneous American and British landings
led by generals George S. Patton and Sir Bernard L. Montgomery
followed by six weeks of ground combat that ended on 17 August, drove
Axis (Italian and German) air, land, and naval forces from the island. It
also opened sea lanes in the Mediterranean for use by the Allies and

helped topple Italian dictator Benito Mussolini from power. On 24 July 1943, the Italian Grand Council of Fascism voted a motion of no confidence against Mussolini. That same day, King Victor Emmanuel III replaced him with Marshal Pietro Badoglio and had him arrested.[14]

Mussolini did not remain in captivity for long. Less than seven weeks later, a group of German paratroopers freed him from the Campo Imperatore Hotel where he was being held high in the Apennine Mountains. The commandos landed nearby from DFS 230 gliders and overwhelmed Mussolini's captors. Otto Skorzeny, a German Waffen-SS officer whom Hitler had personally selected as the field commander for the mission, greeted Mussolini with "Duce [his nickname meaning "the leader"], the Führer has sent me to set you free!" Mussolini replied, "I knew that my friend would not forsake me!" Mussolini was made leader of the Italian Social Republic (a German puppet state in northern Italy). Near the end of the war in late April 1945, with total defeat looming, he fled Milan, where he had been based, and tried to escape to the Swiss border. He was captured and executed near Lake Como by Italian partisans. Afterward, his body was taken to Milan, where it was hung upside down at a service station for public viewing and to provide confirmation of his demise.[15]

Map 9-1

The American Western Naval Task Force assaulted beaches in the Gulf of Gela, and the British Eastern Naval Task Force south of Syracuse

The plan for Operation HUSKY called for the amphibious assault of Sicily by two armies concentrated at its southeastern end, with all landings scheduled for D-day at the same H-hour (exact time when the assault is to take place). British forces including the 1st Canadian Infantry Division were to assault the southern Sicilian coast to the east of Pozzalo, and the east coast, south of Syracuse; while American forces came ashore on the south coast in the Gulf of Gela. The landings would be supported by naval gunfire, tactical bombing, and close air support. General Dwight D. Eisenhower, commander in chief, Allied Forces North Africa, was the overall commander. The British general Sir Harold Alexander was his second in command and the Land Forces/Army Group commander. British admiral Andrew Cunningham was the Allied Naval Force commander. He had under him Vice Adm. Bertram Ramsay, RN, commander, Eastern Naval Task Force, and Vice Adm. Henry K. Hewitt, USN, commander, Western Naval Task Force of the U.S. Eighth Fleet.[16]

OPERATION HUSKY FORCES

Over 3,200 ships, craft, and boats made up the Allied naval forces assembled to launch the invasion of Sicily. The Western Naval Task Force—of more than 1,700 vessels—was charged with landing the American invasion troops on the southwest coast of Sicily at Licata, Gela, and Scoglitti. The soldiers carried by the British Eastern Naval Task Force were to land on beaches on the southeastern side of the island. Lt. Gen. George S. Patton and his Seventh Army would push across the island to secure Palermo, and then swing east to Messina, while British general Sir Bernard Montgomery would drive his Eighth Army north to Syracuse and Catania, to meet up with Patton at Messina. The U.S. Army Air Force and British Royal Air Force were tasked to provide air support.[17]

AMERICAN AND BRITISH HOSPITAL SHIPS

On 11 July, the day after *Acadia* entered port at Algiers, Lt. Col. Thomas B. Protzman learned that she was slated to go into action with eleven British hospital ships and take on patients from the water off the shores of Sicily. The following day, he gave shore leave to all personnel until the 2200 curfew. That evening, British Maj. Gen. Ernest M. Cowell, director, Medical Services, Allied Force Headquarters North Africa, came aboard ship to see Protzman.[18]

He was keen to see *Acadia*'s operating rooms and surgical teams; wanted to know if Protzman and his personnel were prepared to take wounded from the water; asked about readiness of the nurses; and offered some medical advice regarding combat casualties.

> One of his first questions was; "how many Nurses have you got and do you want to take them off the boat before going into action?" I put the question to the girls themselves, and not one of them wanted to quit! He was pleased. I told him of the reprimand I had received from Major General C. P. Gross for putting my Nurses in slacks, and he laughed and said: "You know, the Queen [Queen Elizabeth The Queen Mother] did not approve of my putting British Nurses in trousers while in action, but I still keep on doing it as they are the only clothes they can work in properly. He advised against the use of too many sulfonamides in the burn cases, and due to the shortage of water on Sicily the wounded would be brought in dehydrated and should be given great quantities of water to dilute the sulfonamides they had taken by mouth on the battlefield.[19]

At noon on 13 July, Col. Edgar E. Hume (U.S. Army) and Col. D. Gordon Cheyne (Royal British Army) had dinner with Protzman on board *Acadia*. They were from AMGOT (Allied Military Government of Occupied Territory), which was posed to soon take over the civilian government in Sicily. The fighting was going on so rapidly in Sicily and so few of American soldiers were being wounded that Protzman recorded in his diary:

> We probably won't get up to the front at all. I hate to think of lying out in this harbour (Algiers) with nothing to do. My boys will go nuts, as none of us are allowed off the boat, just waiting several miles from shore. Around 2000 a large convoy of troop transports is coming into the bay escorted by battleships and destroyers. They drop anchor just outside the harbour and will probably supplement the invasion forces or relieve some of the assault troops on Sicily. I sincerely hope we go out with them.[20]

At 0700 the following morning, 14 July (Bastille Day in Algiers), a landing craft came alongside *Acadia* and she took aboard eight wounded British soldiers (5 litter cases and 3 ambulatory patients). Some of them were paratroopers and glider riders who had attacked Syracuse, Sicily, part of the Augusta-Syracuse strong point taken by British forces. Protzman observed about his new charges, "They were pretty badly shot up but happy to get on a Hospital Ship. My girls peppered and spoiled them."[21]

That evening, *Acadia* received orders to stand out to sea and anchor. USAHS *Seminole* had returned to Bone, Algeria, the previous day with wounded men from Bizerte, Tunisia, and *Acadia* was to shortly receive her share. Just as she anchored about five miles from the port area, a Navy boat drew alongside and signaled her to prepare to take on 600 wounded, either tonight, or early in the morning. A fleet of thirty ships was returning from Sicily with casualties and she was to take the ones received to Oran.[22]

Early the following morning, 15 July, *Acadia* again entered port in Algiers, berthing at about 0600 to take on the wounded. Maj. William V. Barney, the ship's chaplain did a splendid job of receiving the wounded; at one point he and his men loaded eleven patients over the sides in three minutes with a single tackle he had devised himself. While other personnel were loading patients (using "Stokes" litters), Brig. Gen. Frederick A. Blesse (chief surgeon, U.S. North African Theatre of Operations) came aboard and was pleased by the unusual sight of a chaplain doing something entirely different from ministerial duties. Blesse told Protzman that during the initial assault, only 680 men were wounded and not a single enemy plane was seen. The men that *Acadia* was then taking on were some casualties of that attack. Her orders were to take on the medical cases from ships landing at Algiers, to full ship's capacity, go to Oran, unload, and await orders there. At 2030, *Acadia* stood out of port for Oran carrying 808 patients.[23]

As she left Algiers, two British hospital ships had been ordered to Sicily as the Italians and the Germans were beginning to fight back and Allied casualties were increasing.[24]

Acadia entered port at Oran at 1100 on 16 July, and berthed at Pier 14. Protzman described the preceding night aboard ship, and other required tasks following her arrival:

> I have been in the OR [operating room] all night picking shrapnel out of people, patching up others, and re-dressing the burn cases. In this short run we're only able to handle the worst and emergency cases and the Hospitals at Oran will continue the work. The doctors and half of my unit worked all through the night as we have to turn complete medical records on the cases as well as take care of them physically. As soon as we reach Oran, we have to unload our patients, take on water, run the laundry for the first time, and wash all the dirty linen before we leave.[25]

Acadia debarked her patients in just four hours at Oran, half of them litter cases. On the 17th, Protzman arranged to have trucks take the nurses and enlisted men out to the beach at Ain-et-Turk in early afternoon for some relief activities. While they were gone, he received an invitation for the nurses to attend a ball in Oran that night. The dance was quite an affair of mixed civilians and military and was held at the City Hall with an orchestra playing. There, Protzman met Brig. Gen. Thomas B. Larkin (commanding general, U.S. Army Services of Supply). When he told him that he was commanding officer of the *Acadia*, Larkin replied, "Yes, I know, you're sailing for Sicily in the morning." Protzman replied, "Maybe, but neither I nor the captain knows anything about this." Upon returning to *Acadia*, word was waiting that she was to sail at 0630 the following morning, but no destination was given.[26]

Map 9-2

Southwest area of the Mediterranean

Acadia put to sea as scheduled on 18 July, hugging the coastline to keep clear of convoy lanes and as far from the Italian base at Sardinia as possible. Her orders were to proceed to Bizerte, Tunisia, which she was expected to reach in mid-afternoon the following day. Early the next morning, about three hours from Bizerte, a convoy was sighted that covered the entire horizon, an amazing and awe-inspiring sight. Shortly after daylight, four groups of Allied bombers, fifty per group, passed over the ship flying in the direction of Italy.[27]

Acadia arrived at Bizerte on the 19th and anchored offshore. Near her were two British hospital ships, HMHS No. 24 and 33 (*Atlantis*). USAHS *Seminole*, which had brought in 200 wounded from Sicily, was alongside a wharf in the harbour. The town was badly damaged from enemy air attacks, and there were many sunken vessels in the harbour. Several Allied submarines passed near *Acadia* as they entered Bizerte harbour, the sky was full of planes, and it was so incredibly hot that men were shirtless, sitting around in their shorts only.[28]

Photo 9-1

Allied anti-aircraft gunfire at Bizerte Harbour, mid-1943.
Naval History and Heritage photograph #NH 123928

On 20 July, all ships were ordered into Lake Bizerte a nearby waterway for safety as an enemy attack of some sort was expected. A French pilot boarded *Acadia*, and she proceeded on her way. The lake lay several miles back from the coast cloistered in the hills, and could only be reached through a narrow channel filled with sunken ships. Protzman described the sight that greeted them upon reaching the saltine, inland body of water:

> The Lake is of considerable size, with salt water, and there must be several thousand ships at anchor here, yet it seems empty. Small Arab villages dot the shores, they appear so white and spotless in the sun, yet, death and misery lurk in every house and tent.[29]

On 21 July, HMHS *Oxfordshire* arrived in the lake and dropped anchor near *Acadia*. Brigadier General Blesse had been in Bizerte the previous day and when he saw five inactive hospital ships, he became quite angry as many wounded were being brought from Sicily in landing craft. He hoped to get *Acadia* out of the lake the following day.[30]

On 28 July, *Acadia* left Lake Bizerte and passing out through the channel, she found the outer harbour was filled with assault landing craft of all descriptions loaded with troops and equipment ready for combat. Later that day or early the next she put to sea.[31]

Photo 9-2

Troop boarding of infantry landing craft at Bizerte, Tunisia, on 6 July 1943. National Archives photograph #Sc 176486

At 0600 on the 29th, *Acadia* was off Sicily, which appeared to her crew and medical staff to be just as mountainous and foggy as the coast of North Africa had. A small destroyer led her through the minefields off Palermo via the swept channel, and into the harbour. The Port Officer ordered the hospital ship to anchor 1,400 yards from shore. As soon as she had done so, another destroyer came alongside to embark wounded who had been involved in land combat the previous evening.[32]

Protzman then went ashore by boat, accompanied by some others for reasons explained in a diary entry:

> I had one of my Officers with me and two of the girls (I find it much easier to get by any guards into a restricted or forbidden area if we take a couple of the best-looking Nurses with us).[33]

Reporting in at the Port Director's Office, Protzman and company were directed to Seventh U.S. Army Headquarters in the Post Office building in the city. There, Col. Daniel Franklin (Surgeon, Seventh United States Army) told Protzman that he had been trying for the past three days to get the *Acadia* over to Palermo and clear the wounded out of the field hospitals, so they would be able to care for those that would be wounded in the coming battles. Patients were being evacuated from the front, and *Acadia* was to load them from landing barges, because the Port Director didn't feel that it was safe for a ship of *Acadia*'s type and size to remain in the harbour.[34]

While *Acadia* did not receive any patients that day, at 0800 the next morning, 30 July, there were still no landing barges in sight. That soon changed. As many as three landing craft were alongside her at a time, filled with patients brought in directly from the front, while the chaplain was taking on the litter cases at less than a minute for each one. Protzman observed this about the general condition of patients: "The medical staff in the Battalion Aid Stations and the men and women in the Field Hospitals are doing a wonderful job, as the men who are coming in have all had excellent care."[35]

As landing craft came alongside, *Acadia* shock teams went aboard them and distributed coffee and drinks to those that didn't have stomach or chest wounds, and sent the worst wounded on board the hospital ship to receive immediate surgical attention and care.[36]

In one of the barges were a lot of wounded French Moroccan Goumiers ("Goums"). Recruited from the hill tribes of Morocco's Atlas Mountains, the knife-wielding irregular troops distinguished themselves fighting under French command in Tunisia, Italy, France and Germany during World War II. Garbed in the traditional djellaba of their homeland, and armed with long sharp knives, in addition to rifles, machine guns and mortars, their ferocity terrified the enemy. The Goums skill in mountain warfare had prompted Gen. George S. Patton to request their participation in his Sicilian campaign, where they fought brilliantly in it and many other key campaigns.[37]

Most of the Goums spoke a little French. When Protzman asked one of them why they wore crude homemade sandals instead of Army boots, he said it was impossible to creep up on a man and cut his throat without him being warned if they wore the heavy shoes. Their main weapon was a heavy knife with a razor edge that they could equally well shave with as chop off an enemy's head.[38]

By dusk on 30 July, *Acadia* had loaded nearly 500 patients, all that were then available to bring on board. The medical staff was advised a hospital train was expected from the front the following day, if the tracks could be repaired. This must have occurred, because shortly after noon on the 31st, landing craft began to arrive at her side. By dusk, *Acadia* had loaded many more, nearly a thousand patients in total. On the hospital ship's exit from Palermo, she had to anchor clear of the channel for a short time, to allow passage for several troop transports that were carrying 25,000 newly arriving soldiers inbound to join the fighting ashore.[39]

That night while *Acadia* was out in the Mediterranean en route to Bizerte, Tunisia, Protzman described the efforts of his surgeons, nurses, and other medical personnel in providing patients' necessary care. *Acadia* travelled with her lights on (identifying her as a hospital ship):

> We all worked through the night doing dressings or operating. It was so hot in the OR that my shoes filled with sweat at the end of each operation. We finally discarded the white operating gowns and stripped to the waist dressed only in sterile gloves and trousers. Most of the cases were badly infected so it didn't really make any difference what we wore. We lost one man during the night with a bad shrapnel wound in the leg and thigh, because he had gas gangrene.... We have a great part of the work cleaned up and will probably be in bed by 0100 or 0200 next morning.

> The Nurses have done a fantastic job. All of the wounded have been bathed and cleaned; their dressings changed or made ready for the Operating Room. We now return to Bizerte, Tunisia, to fill Jack Harowitz' 53d Station Hospital to overflowing![40]

In the evening on 1 August, *Acadia* berthed alongside the pier in Bizerte and discharged all of her patients in two and a half hours; 551 of which were litter cases and the rest ambulatory. Over one-third of them had malaria. Needing more Quinine, Protzman sent some men to the 7th Medical Supply Depot, noting about the preferable use of this medicine over an alternative and the condition of some of the wounded personnel:

> We use it more often as Atabrine is not that good for treatment since it causes diarrhea and increases suffering. Over 300 of the boys have sand flea fever, and quite many of them have severe burns from tank battles.[41]

At 1000 the next morning, 2 August, USAHS *Seminole* and HMHS *Lancashire* departed for Sicily. This left HMHS *Oxfordshire*, HMHS *St. Andrews*, and USAHS *Acadia* at Bizerte. Protzman believed that *Acadia* would be in port for some time, recording in his diary the following day:

It sure looks as though we're going to be stuck here again for some time and I will have to figure out something for the Nurses and the EM [enlisted men] in the form of diversion or recreation as it has been some time since any of them have been ashore. Transportation is difficult to get as all the landing craft we used on the last trip here are at the front and it may just be impossible to send the boys and girls ashore. I've called in the ARC [American Red Cross] ladies and asked them to stage another dance on the top deck and maybe among the group of Officers they invite I might find one with an LCT [tank landing craft] or LCVP ["Higgins boat"] to take my Enlisted Men to visit Ferryville. [Today called Menzel Bourguiba, this town is in the extreme north of Tunisia]. I know a good many men will probably get drunk, but it will help release a lot of steam.[42]

Protzman noted about social activities aboard *Acadia* on 3 August, involving hosting their British counterparts to watch a movie, and the afore mentioned planned dance:

We had the entire ship's company of HMHS *St. Andrews* aboard to watch a movie. They enjoyed it a great deal as they hadn't seen one for the last eight months, while we have one almost every other night. The dance was a success and the girls will be happy for a few more days.[43]

On 5 August, Protzman visited the Port Headquarters and found out that *Acadia* was to be loaded with patients for transport to Oran, as the local hospitals were overcrowded. At 0630 the following morning, she received her first batch of wounded from the 56th Evacuation Hospital. They and succeeding patients from this hospital totaling 786 wounded, took a little over two hours to embark. The final patients coming from the 3d General Hospital were loaded in exactly one hour. Protzman noted this about the wounded personnel on board:

Our cases are so interesting and a great many are British. One little Limey boy lies in bed with his body in a plaster cast from the waist down; he proudly shows me the piece of shrapnel that put him out of action (about 3 inches in diameter). The shell fragment went through his right hip, came out through his bladder, and lodged in his left leg, from where it was removed. He has also six MG [machine gun] bullets in his back and may well be crippled for life. He is a brave kid and rarely complains.[44]

On 7 August as directed, *Acadia* sailed from Bizerte. Protzman noted in his diary conditions at sea that day, and at the church service the following morning before the ship's arrival at Oran at 1100:

> This is the first time we've been out in the Mediterranean, that it has been so quiet and peaceful. The whole day has been uneventful.
> The church service was a record one today and the deck was crowded with the patients able to reach the area. It's pitiful how these young homesick wounded kids come to the service for mental and spiritual aid. They are all reaching out to that intangible 'something' we call God, and it's amazing what comfort they get from a few songs and a prayer.[45]

On 9 August, Acadia moved from Oran to nearby Mers-el-Kebir in early morning to make room for another ship. That evening she received orders to sail for Bizerte at daylight. The next day, following the North African shoreline to Bizerte, there was a noticeable difference from past transits. The entire coast was being patrolled by small craft, suggesting that something had occurred to put the Navy on the alert.[46]

Acadia anchored off Bizerte on 11 August. A port officer came aboard to inform Protzman that they hoped to load the hospital ship two days henceforth, and send her back to Oran. Protzman noted about conditions at that Tunisian port city:

> Last week, Bizerte suffered one of the heaviest bombardments in its history shortly after we left port, and two British destroyers were hit. The outer harbour is full of LSTs [tank landing ships] jammed with PWs [prisoners of war] and just before darkness set in, a long line of boats [ships] started on a convoy bound for Sicily.[47]

At noon the following day, 12 August, following an uneventful night for those aboard ship, *Acadia* started down the narrow channel to Lake Bizerte. Just as she passed the pier headquarters the red alert flag went up and a siren started wailing. Anti-aircraft artillery started shooting. The Arabs on the shore started running to the shelters, and the soldiers to man their guns in the face of an air raid by the German Luftwaffe, which Protzman described:

> A British battleship opened up with her guns, which set our ears to humming. I had a hard time ordering everyone to put on their helmets as they were so obsessed with watching the fight. Most of the injuries sustained in an air attack, outside of direct bomb hits, are from the falling pieces of shell fragments. Several bombs fell on

Ferryville, on the shore near us, but very little damage was done, and one of our fighters shot down an enemy bomber.[48]

At 2200 that night, HMHS *Oxfordshire* arrived in port from Sicily, but was ordered back out to sea as another air attack was expected. She was scheduled to reenter port later. *Acadia* received orders to prepare to come alongside the pier the following morning after *Oxfordshire* debarked her wounded.[49]

Acadia's expected 13 August patient embarkation did not materialize. Protzman explained the reasons for the delay and resultant effect on the parched wounded waiting to board the ship:

> As usual, the harbour got things gummed up. The Lake was full of ships now. At 1530 a pilot appeared and we transferred to a loading dock in Bizerte, unfortunately the wind and the current were so bad that it was almost 1800 before we were safely anchored alongside and taking on patients. The poor devils had been waiting in the heat for over four hours and had had no food. Of the 800, 250 were NP [neuro-psychiatric] patients; they marched onto the boat staring straight ahead, with little or no expression on their faces, yet the slightest disturbance throws them into a panic that is sometimes difficult to control. After we left the harbour, the ship was anchored for two hours awaiting further orders, and just this waiting and the change in movement of the vessel set most of them on edge. What they feared most was an air attack, and the last one yesterday didn't really help their condition.[50]

Leaving Bizerte, the sea conditions in the Mediterranean were ideal and with a following wind, *Acadia* expected to arrive at Oran shortly after midday on 15 August. Protzman noted, "Our radio picked up the position of the *Seminole* and from her location she is also on the way from Bizerte to Oran. This means that all our hospitals in the advance are overflowing."[51]

Acadia reached the pier at Mers-el-Kebir at 1530 on the 15th and that evening began to unload her wounded. All of her patients were unloaded in good order, and medical staff were relieved to get the NP cases into better hands, where they would receive proper treatment. The *Seminole* arrived in port on 16 August to be loaded with patients for a return trip to the United States to help relieve the overcrowded situation in hospitals in North Africa and to be fitted out properly so that she could perform winter service.[52]

On 17 August, *Acadia* received word that the *Seminole* had begun her voyage to the United States with 537 patients on board. At 1300 that day, *Acadia* was proceeding to sea, en route to Bizerte to relieve the congestion in that area. Protzman observed about the ship's schedule, "We had hoped to return to Sicily, but will have to go where the pressure is the greatest. Where will we go from here I don't know, as all Hospitals in the region are full. Possibly Algiers?"[53]

This is a good point to end this chapter devoted to the service of USAHS *Acadia* and USAHS *Seminole* during the invasion of Sicily. As summarised in the remaining page and a half, the city of Messina fell to the Allies on 17 August, marking an Allied victory and successful conclusion to Operation HUSKY.

CONCLUSION OF OPERATION HUSKY

The capture of Sicily was an undertaking of the first magnitude.

—British Prime Minister Winston Churchill.

The brilliant achievements of the Allied forces in this conquest, launched on a magnitude which heretofore had never been attempted, were due principally to the singleness of purpose which all forces demonstrated. The appreciation of each other's problems produced an inter-service spirit of co-operation and common endeavour which welded the naval and military forces into a single team possessed with the resolute will to win.

—Vice Adm. Henry K. Hewitt, USN, commander, Western Naval Task Force.[54]

General Patton and his staff left Admiral Hewitt's fleet flagship *Monrovia* (APA-31) at 1700 on 12 July, and soon after leaped from a landing barge and waded ashore to the beachhead at Gela. His troops—consolidated from the three landing positions—advanced across the island to Palermo, then fought their way along the north coast, aided by the Navy. On 27 July, when Palermo Harbour was first opened to Allied shipping, Vice Adm. Henry K. Hewitt organised Motor Torpedo Squadron 15, Destroyer Squadron 8, several minesweepers, and a few other warships, left in Sicilian waters, into "General Patton's Navy," to support the Seventh Army.[55]

Photo 9-3

British Gen. Bernard L. Montgomery (centre) and
Lt. Gen. George S. Patton Jr. look over a map of Sicily, circa July-August 1943.
Naval History and Heritage Command photograph NH-95561

This force was commanded by Rear Adm. Lyal A. Davidson, embarked in the cruiser *Philadelphia* (CL-41). He arrived at Palermo on 30 July with the cruiser *Savannah* (CL-42) and destroyers *Butler* (DD-636), *Cowie* (DD-632), *Glennon* (DD-620), *Herndon* (DD-638), and *Shubrick* (DD-639). After anchoring outside the breakwater, Davidson met with Patton to discuss operations along the north coast of Sicily. These would include gunfire support for the Seventh Army as it advanced, providing craft to Patton for shore-to-shore amphibious operations to "leapfrog" enemy strongholds, and ferrying Army heavy artillery, supplies, and vehicles.[56]

On the night of 16 August 1943, beating the British, the leading elements of the 3rd Infantry Division entered Messina, and the city fell the next day. In just thirty-eight days, the Seventh Army, under Patton's leadership, and the British Eighth Army, under Montgomery, conquered all of Sicily. Thus ended Operation HUSKY, which Patton observed "out-blitzed the inventors of Blitzkrieg."[57]

10

British Hospital Ship *Talamba*

13.7.1943.
Tuesday.
British Hospital Ship TALAMBA *was deliberately attacked and sunk by enemy aircraft whilst embarking casualties off Avola night 11/12 July. Ship fully illuminated. Casualties unknown.*

> —Admiralty War Diaries entry of 13 July 1943.
> *Talamba* was actually sunk on 10 July 1943.

D Day. 10th July
2155 H.M. Hospital ship TALAMBA, *lying some three miles south east of the anchorage and showing lights was bombed: This attack was deliberate.* TARTAR *was the first ship to close the* TALAMBA *and did very good rescue service.*
2300 UGANDA *closed at about 2300 just before* TALAMBA *sank.* CARLISLE *came next and was allowed to help rescue survivors, but* MAURITIUS *was sent away as too many ships were concentrating.* UGANDA *picked up 75, which included some casualties which had come straight out from the beaches looking for* TALAMBA. *The ship was thus saddled with a large number of casualties for whose accommodation it was necessary to the use the Ward Room, Gun Room, Warrant Officers Mess, E.R.A's Mess and eventually a hangar.*

One was brought on board dead. One died on board. These were buried at sea.

With the survivors were two Army Doctors, a Chaplain and a Nurse. These did great work among the wounded, many of whom had Received nothing but first aid before arrival on board.

D Day + 1. 11th July
0200 By 0200 the work of picking up survivors was complete and UGANDA *was kept under weigh in the vicinity of the convoy anchorage.*

> —From The Report of the Commander-in-Chief, Mediterranean on the Invasion of Sicily (Operation "Husky").

Photo 10-1

Hospital ship HMHS *Talamba* sailing into harbour in World War II.
Photograph from the collections of the Imperial War Museums

SS *Talamba*, a 450-foot, 8,018-ton passenger/cargo ship, was built in 1924, by R. & W. Hawthorn, Leslie & Co Ltd, Hebburn Yard, Newcastle, UK. Propelled by steam turbines coupled to twin propellers, she could make 16 knots. Her third funnel was a dummy, built to impress potential passengers. It suggested that she had a third fireroom, additional boilers, and thus greater speed.[1]

She was employed by the British India Steam Navigation Co Ltd, London, which named their ships after places in Asia, particularly the Indian sub-continent; *Talamba*'s namesake was a town near Multan in Pakistan. She mainly sailed between Calcutta, Rangoon, Penang and Singapore—offering accommodations for 56 1st-class, 80 2nd-class, and 2,777 deck passengers—with occasional voyages further north to China and Japan during this service.[2]

In 1939, she was used as a troop ship between India and the United Kingdom. *Talamba* became a hospital ship in June 1941 and was based at Colombo, Ceylon (today Sri Lanka). In January 1942, she assisted in the evacuation of Singapore.[3]

Map 10-1

Sri Lanka (formerly Celon) is a South Asia island country in the Indian Ocean, southwest of the Bay of Bengal

Photo 10-2

Drawing by Frank Norton of barrage balloons flying (to defend against dive bombers) above Royal Australian Navy ships HMAS *Lismore*, *Cairns*, *Launceston*, and *Ipswich* in Colombo Harbour, Ceylon, 1944.
Australian War Memorial photograph ART21097

Despite being clearly marked as a hospital ship, *Talamba* was sunk off Syracuse on 10 July 1943, by Italian aircraft attack during the Allied invasion of Sicily. The following account of her loss is gleaned from an article written by Stan Fernando, a Junior Engineer from Sri Lanka, who served on board *Talamba* at Sicily.

The *Talamba* was a happy ship with a full complement of medical staff including nurses on board, and practically all the ships officers had sweethearts and girlfriends which made life worth living in spite of the dangers existing at sea in war time. Fernando described the principal ship's officers, and paid tribute to her Indian crew who conscientiously carried out their important duties amidst the dangers of war:

> The Captain was an Englishman by the name of J. D. Woods who lived up to the best traditions of his profession. Brodie a dour wee Scotchman was the Chief Engineer under whose special jurisdiction I came, while Olson a typical Geordie, was the Second Engineer and I as Junior Engineer kept the 4-8 watch with him. Big Aussie Brown was the Third Engineer and a braver man I cannot think of, for it was he who volunteered to go down the Engine Room and put out the generators when we realized that the lights were attracting the aggressive plane. Of course Aussie Brown had extra reason to be careful as he was engaged to be married to one of the lovely nurses on the medical staff on board.

We had an Indian crew both on deck and in the Engine Room. Right through the watch the firemen who kept the boilers fired and the greasers (oilers) went about their work quietly and efficiently, not appearing in the least bit perturbed by the fact that all around us there was the constant sound of bombs bursting. It would be no exaggeration to say that the British Merchant Navy owed a great debt to the Indian seamen who I would say manned at least 50% of the British Merchant ships.[4]

Talamba left Tripoli, Libya, the night of 9 July 1943 and joined a large ship convoy northbound for Sicily. No one, including the ship's captain, knew their destination prior to departure. Woods had been given sealed orders to be opened only after leaving port.[5]

Map 10-2

South-central area of the Mediterranean

At dawn on 10 July, crew or medical staff up on deck could see that *Talamba* was at the back of a vast convoy of ships stretching as far as the eye could see. It was only then that those on board realised that they were headed for Syracuse in Sicily, and this was the beginning of the assault for its capture. Upon learning this, Fernando was not pleased that she had been "blacked out" during the night—the explanation being, of course, that utmost secrecy had to be maintained.[6]

That night off Syracuse, *Talamba* was flood-lit to ensure that no mistake would be made in the identity of the ship's mission. Fernando described finishing his engine room watch at 2000, and looking forward to some rest. His allusion to manoeuvring of engines in the following quoted material refers to almost continuous adjustment of the ahead and astern guarding valves sending steam to the turbines, to answer "bell orders" from the bridge, necessary to maintain station in the convoy.

> After a gruelling four hours of watch keeping with constant manoeuvring of engines, I was glad to come up when relieved by the Fourth Engineer and his mate at 8pm. This was the high moment of sea watches for me – to come out of a hot Engine Room after four hours of sweating it out and then to know the comfort of a cool bath and sliding between clean sheets with a good book in hand and reading yourself to sleep and oblivion.[7]

Being very tired, Fernando immediately drifted off to sleep before being awakened a short time later by an Italian aircraft attack involving bombing and strafing of the ship, followed by orders from the bridge to extinguish all lights. The lighting illuminating and clearly identifying *Talamba* as a hospital ship, also made her an easy target for enemy pilots ignoring International laws regarding the sanctuary of hospital ships:

> I must have just fallen asleep when there was a terrific bang and woosh! A great splash of water hit the boat deck, I was out of my bunk in a flash and put on a boiler suit. In a trice I was outside my cabin at the same time as everyone else; there appeared to be no sign of any damage but the boat deck was awash. We heard the screech of an aero-plane race past with machine guns blasting away and there came a booming voice saying 'put those lights out'….

> In a matter of minutes we were aware that the *Talamba* had reached her doom. The vessel was now sinking quite quickly and word went round to lower the lifeboats. Aussie Brown had gone down below and put out the power supply; cutting out the lights. No sooner we knew it was 'abandon ship' than everyone put on his life jacket, popularly known as 'May West' in those days. In no time the life boat I was in charge of was lowered with assistance of search lights from ships around us.[8]

Fernando declined to enter the life boat for which he was in charge (nor did most of the other officers regarding their boats). After seeing the boat launched and rowed off, he instead went around the ship trying to be of assistance, but found there was not much he could do because:

> The 'aft' end of the ship was now sinking and everybody left on board had wandered to this end where the main deck was practically awash. A great sense of camaraderie existed among us ship mates and this was further enhanced by the unfortunate circumstances borne by all of us. But there was no time left to help each other anymore.

It was now about twenty minutes after the bomb had exploded and time was running out. Fortunately we were surrounded by other ships which gave us a great sense of security.[9]

Walking along the upper deck, Fernando, who had never learned to swim, came across a solitary life buoy. Picking it up, he leapt from the ship into the water along with several others. Seeing, in the blaze of other ships' searchlights, *Talamba* sinking, and remembering that a ship when going down sucks everything within range with it, he started to get scared—particularly remembering his widowed mother back home in Ceylon and prayed for God to save him at least for her sake.[10]

In the water, Fernando could hear others shouting for help as in great panic, he was trying to get as much distance as he could between himself and the doomed ship. When he looked back in fear it was to see the most awesome sight of his life captured in the glow of many searchlights of the surrounding ships:

> The sight of this once majestic ship literally standing upright with its bows pointing skywards and all three funnels unbelievably horizontal, one by one the funnels broke adrift and tumbled into the sea and the vessel herself dramatically sank below the water surface taking with her the few hundred doomed and desperate souls [wounded personnel taken on board earlier in the day before the aircraft attack] to their watery grave.[11]

Enemy aircraft attacks had begun at 1550 that afternoon (D-Day) with an air raid on 'Acid' landing beaches and landing craft by Ju 88s and fighter bombers. Around 2300 that night, *Talamba*, lying some three miles southeast of the anchorage and showing lights was bombed in a deliberate attack. Nearby ships recovered survivors as described in a Commander-in-Chief Mediterranean report:

> [The destroyer HMS] *Tartar* was the first ship to close the *Talamba* and did very good rescue service. [Light cruiser HMS] *Uganda* closed at about 2300 just before *Talamba* sank. [Light cruiser HMS] *Carlisle* came next and was allowed to help rescue survivors, but [Light cruiser HMS] *Mauritius* was sent away as too many ships were concentrating. *Uganda* picked up 73, which included some casualties which had come straight out from the beaches looking for *Talamba*.... With the survivors were two Army Doctors, a Chaplain and a Nurse. These did great work among the wounded, many of whom has received nothing but first aid before arrival on board.[12]

11

Operation AVALANCHE

In the early phases of the landings it became apparent that the enterprise was a gamble, and that the narrowest of margins would govern its success. The German artillery, armoured force, infantry and air force met the Fifth Army at the shore line. On some beaches the early waves were permitted to land and were then pinned down on the narrow beaches by artillery, mortars and machine guns which also took under barrage fire succeeding waves of landing craft bringing ashore later waves of troops and equipment. The well-placed strong points, which the Germans had established behind the beaches during the fortnight preceding the landing, brought a withering fire against the soldiers debarking from boats and craft, and the numerous formidable "Tiger" tanks, deployed to cover the strong points, greatly increased the scale of opposition thrown against the incoming waves.... The weight of sea power determined the issue.

—Vice Adm. Henry K. Hewitt, commander Eighth Fleet
and Western Naval Task Force, describing how it was
"touch and go" whether a foothold would be established
on the Italian mainland during Allied landings at Salerno.[1]

Only twenty-three days elapsed between the fall of Messina on 17 August 1943—marking the end of the Sicily campaign—and the commencement of Allied attacks against Axis forces in the Gulf of Salerno on 9 September. More than six hundred Allied warships, other type ships, and landing craft participated in the amphibious invasion at Salerno. Many of the ships and most of the landing craft had taken part in the Sicily campaign over the previous two months. Limited port capacities on Sicily made it necessary to disperse these forces over virtually all serviceable Mediterranean ports from Oran, Algeria, to Alexandria, Egypt, during the planning and training phases preceding the Salerno landings. This prerequisite presented challenges to command echelons of all services regarding the coordination of an already most difficult operation.[2]

After the completion of a short preparatory period, the Western Naval Task Force transported Fifth Army forces to an attack position off Salerno for a pre-dawn assault over beaches against well-established German forces. The Gulf of Salerno was chosen for the landing on the Italian mainland for two important reasons. It was located only about forty miles south of Naples, a major port to the south of Rome, which the Allies wanted to capture for use as a supply base. Additionally, although Salerno lay two hundred miles north of Sicily, it was within range of fighter-plane support operating from Sicilian airfields.[3]

ORGANISATION AND GEOGRAPHY

The goal of the British and American Corps, which comprised Fifth Army under American Lt. Gen. Mark W. Clark, US Army, was to knock Italy out of the war and shatter the Axis coalition. The overall commanders for the invasion of Salerno, code named Operation AVALANCHE, were Gen. Dwight Eisenhower, Adm. Andrew Cunningham, Gen. Harold Alexander, and Air Chief Marshal Arthur Tedder, the same as in Sicily. Vice Admiral Hewitt, USN, commanded the amphibious force, which was divided into a combined Northern Attack Force—mainly British, under Commodore Geoffrey N. Oliver, RN, and Rear Adm. Richard L. Conolly, USN—and a Southern Attack Force—mainly American, commanded by Rear Adm. John L. Hall Jr., USN. The Gulf of Salerno was correspondingly split into a Northern attack area on which the Northern Attack Force would land X Corps, British Army; and a Southern attack area on which VI Corps, U.S. Army, would come ashore.[4]

The British Northern attack area was divided into "Uncle," "Sugar" and "Roger" sectors, relating to same-named assault forces and, within each of the sectors, the assault beaches. Stretching from farthest north, southward along the coast, were the Red beach and Green beach of Uncle sector, Amber beach and Green beach of Sugar sector, and Amber beach and Green beach of Roger sector. The Uncle and Sugar sectors lay to the south of Salerno Town between the Picentino River (to the north) and the Tusciano River (to the south). Roger Sector was southward of the river Tusciano. The single sector in the American Southern attack area lay in front of the city of Paestum. Renowned for its well-preserved ancient Greek temples, Paestum sat atop the bluffs above the assault beaches. The four beaches below it were designated from north to south, red, green, yellow, and blue.[5]

In supporting actions, U.S. Ranger and British Commando Brigades were to land on the northern shore of the Gulf of Salerno on D-Day; the 1st, 3rd, and 9th Rangers at Maiori beach, and the No. 1 and

No. 41 Royal Marine Commandos at Vietri sul Mare beach.   The Rangers had orders to secure Maiori and destroy nearby coastal defences before moving inland to capture Chiunzi Pass, near Salerno, to control the road to Naples. The commandos were to seize Salerno.[6]

Map 11-1

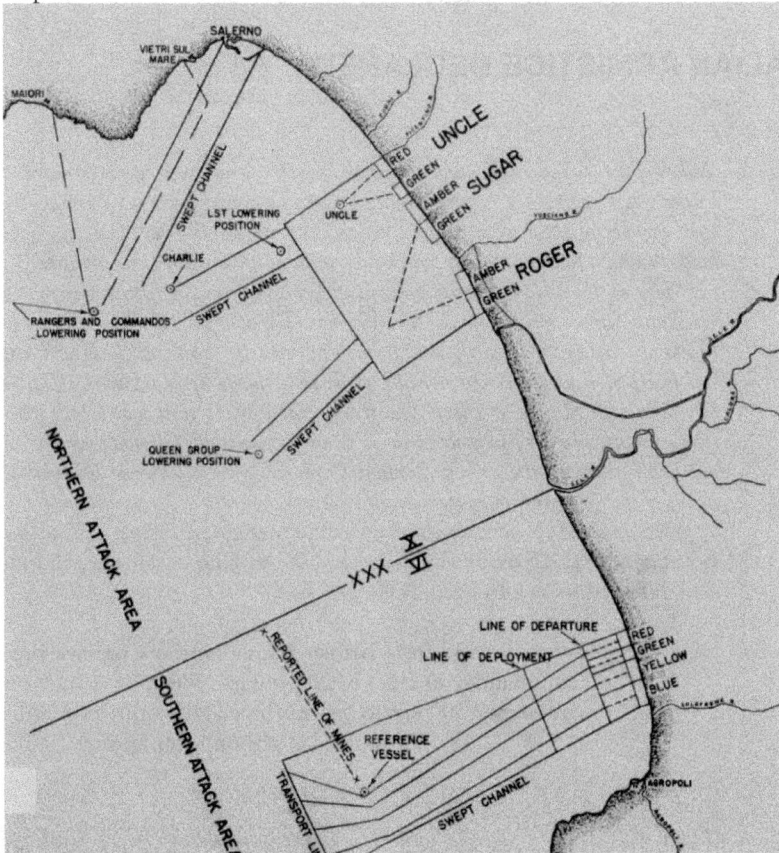

Salerno assault beaches

There were inherent shortcomings to the plan. In order to land two Army corps abreast, it was necessary to come ashore between Salerno and Paestum (to the south) because it offered the only beach area broad enough. However, the Sele River bisected the landing area, which would split the two assault corps. Top Army brass were well aware of this weakness. They knew the Germans could exploit it, slice the beachhead into two segments, and mass greater force against one Army corps at a time. Lt. Gen. George Patton wrote in his diary, following a brief on the operation, "As sure as God lives, the Germans will attack down that river." Patton was then "in exile" in Sicily, having slapped a soldier whom he believed to be a malingerer. Following this well-publicised episode, wide-spread demands for Patton's removal from command, in Congress and in the press, ensued. General Eisenhower condemned Patton's actions, but asserted that his military qualifications, loyalty, and tenacity made him invaluable in the field.[7]

ITALIAN ARMISTICE DECLARED

Sailors of the Italian Navy and Mercantile Marines: Your country has terminated Hitler's war against the United Nations. German armed forces have become the open enemies of the Italian people, whom they have so often betrayed.... Take heed; therefore, that you do not scuttle your ships or allow them to be captured....

Ships in the Mediterranean sail to a place safe from the interference of the Germans. Sail, if you can, to North Africa or Gibraltar, to Tripoli or Malta, to Haifa or Algiers, or to Sicily, and await final success. Ships in the Black Sea sail to Russian ports. Ships not able to do this, sail to neutral ports.... Those who are in the Aegean or Black Sea, if you cannot make good your escape from the Germans, who are now your enemies, do not let your ships fall into their hands. As a last resort, scuttle them or sabotage them, rather than let them fall into the hands of the Germans, to be used against Italy....

Ships intending to act in conformity with this message may assure safe conduct by calling Malta, Algiers, or Alexandria on 500 kcs [kilo cycles per second, now referred to as kilo hertz (kHz)].

—Radio message from British Adm. Andrew Cunningham,
commander-in-chief Mediterranean Fleet, broadcast on
frequency 500 kcs to Italian ships at fifty-three minutes
past midnight on 9 September 1943.[8]

Vice Admiral Hewitt was informed by the Allied Northwest African Tactical Air Force, while en route to the Salerno landing beaches, that an Italian armistice had been declared and that aircraft encountered bearing Italian markings were not to be attacked, unless they demonstrated hostile intent. Similar reports and guidance from other Allied commanders followed. Following a radio broadcast by Admiral Cunningham, commander-in-chief Mediterranean Fleet, to Italian ships and submarines, providing guidance regarding their surrender, Hewitt instructed Western Naval Force units that operations in progress were vigourously proceeding as ordered, and that:

Italian armed forces, including aircraft, should be treated as friendly unless they take hostile action or threaten hostile action. Plans for covering fire on beaches are to progress as ordered, but coast defence batteries should not be engaged unless they open fire.[9]

Italian forces on Ventotene Island, one of the Pontine Islands in the Tyrrhenian Sea that lay twenty-five nautical miles off the coast of Gaeta, Italy, had surrendered at 0001 on 9 September without resistance. However, for Allied forces approaching Salerno beaches and landing troops soon thereafter on D-Day, it probably made little difference whether German or Italian forces might be opposing them. At 0215, shore batteries on the beach opened fire.[10]

BRITISH NORTHERN ATTACK AREA

The decision to delay the pre-assault bombardment in an effort to maintain the element of surprise proved to be a costly one. Personnel on the Uncle Attack beaches were pinned down by enemy 88mm fire, which should have been neutralized or knocked out by adequate naval gunfire prior to the landing. During the final approach to the initial transport area, commencing at 2100/D-1, enemy air attacks indicated that surprise had been lost.

—Vice Adm. Henry K. Hewitt, commander Western Naval Task Force, commenting on the folly of forsaking shore bombardment to soften German defence positions, and launching the assault on Salerno beaches in darkness— which U.S. Army leadership insisted upon— in the hope of achieving surprise.[11]

The mission of the Northern Attack Force was to establish the 46th Division of British X Corps ashore over Uncle sector's Red and Green beaches, about five miles south of Salerno at H-Hour on D-Day 0330 on 9 September. The landing plan called for a one brigade front carried in LSTs, LCI(L)s, and LCTs. Six of the tank landing ships in the assault group carried pontoon causeways to ensure unloading of LSTs over beaches expected to be too steep for effective use of ships' bow ramps. Lesser gradients were encountered, and the causeways proved unnecessary—the landing ships and craft beached "dry-shod" to transfer materials ashore.[12]

Enemy fire began with the landing of the first boat wave. Coastal batteries on adjacent hills and mobile batteries located about one mile inland opened fire on the Uncle beaches. British assault troops were forced to abandon Green beach due to heavy enemy shelling. By 1240 on D-Day, the enemy had overrun it and was threatening the right flank of Red beach. Allied reinforcements (one company of the 5th Foresters and one company of the 5th Leicesters) were landed by LCI(L)s on Red beach, and concentrated naval gunfire from ships offshore was brought to bear against enemy targets. Red beach was consolidated (brought under Allied control), about daybreak on D-Day, but quickly became congested due to the arrival of craft diverted from Green beach. Despite limited beach space to unload and ongoing enemy bombardment and aircraft raids, fair progress was being made by noontime on D-Day.[13]

Roger and Sugar sectors received moderate fire from enemy guns during the early hours of the assault, but naval gunfire quickly neutralized enemy defence positions. Tank landing ships were brought in after daylight with smoke employed to cover their approach. Although X Corps was under enemy fire until 26 September, the beaches of British Uncle, Sugar, and Roger sectors were brought to order earlier than the American beaches south of the Sele River.[14]

AMERICAN SOUTHERN ATTACK AREA

Rear Adm. John L. Hall Jr., commander Eighth Amphibious Force, was in charge of the Southern Attack Force (Task Force 81). This force was responsible for establishing the U.S. Army 36th Infantry Division, and attached troops of the VI Corps, ashore over beaches south of the Sele River. The various elements of the force sailed from Oran and Algiers, Algeria; Bizerte, Tunisia; and Palermo, Sicily, in three convoys and rendezvoused off Salerno around sunset on 8 September. After forming an approach disposition, the force proceeded to the assault area.[15]

FIERCE COMBAT ASHORE

Salerno beachhead ranks with those of Anzio, Tarawa and Normandy as the most fiercely contested in World War II. Few soldiers came under such severe fire on landing as did those of the American VI and the British X Corps, or came through it so well. Yet even these "valiant men of might" could not have carried on without naval gunfire. For three days, 10-12 September, as the tide of battle swept to and fro over the Salerno plain, both navies delivered gunfire support to the troops ashore. During this period the Luftwaffe attacked ships in the roadstead with a new glide bomb, radio-directed to its target from a high-flying plane.

—Samuel Eliot Morison in *The Two-Ocean War.*[16]

Following herculean efforts by Allied minesweeping and other forces to facilitate landing Allied assault forces ashore at Salerno, British and American troops after landing found themselves immersed in fierce land warfare. General Heinrich Vietinghoff, commander of the German Tenth Army, had three divisions assembled on the Salerno plain with 600 tanks and mobile guns. With this powerful force, he was posed to throw Lt. Gen. Mark W. Clark's Fifth Army into the sea before General Montgomery could come up and relieve him. (Monty's Eighth Army's XIII Corps had crossed the Straits of Messina on 3 September and landed 300 miles south of Salerno at Calabria.)[17]

Vietinghoff's intention was thwarted by unrelenting naval shore bombardment from the sea. He reported to Field Marshal Albrecht Kesselring (senior German commander in Italy) on 14 September, "The attack this morning…had to endure naval gunfire from at least 16 to 18 battleships, cruisers and large destroyers… With astonishing precision and freedom of manoeuver, these ships shot at every recognised target with overwhelming effect."[18]

The next day, 15 September, German armoured units made a final attempt to dislodge the Allies from the Salerno beachhead. It was unsuccessful, and the day after, Kesselring ordered a general retirement, "in order to evade the effectiveness of shelling from warships," as he explained in his memoirs.[19]

U.S. ARMY HOSPITAL SHIPS UNAVAILABLE

During the Allied battle for the Salerno beachhead, 9-17 September 1943, there were not any American hospital ships available to transport and care for wounded personnel. USAHS *Seminole* had left Mers-el-Kebir, Tunisia, on 17 August, with 537 patients on board, beginning her return voyage to the United States for shipyard modifications. USAHS *Shamrock*, the U.S. Army's third hospital ship to be put into service in World War II, arrived at Mers-el-Kebir on 18 September. Possessing a 655-bed capacity and a top speed of 14 knots, she sailed the very same day for Palermo in Sicily. More about her later.[20]

ACADIA OCCUPIED AWAITING DRYDOCK REPAIRS

On 1 September 1943, *Acadia* was laying in the outer harbour at Oran, Algeria, near a torpedoed freighter—pier space being at a premium. This was the largest concentration of troop and supply ships ever congregated in the harbours of Oran and Mers-el-Kebir; and this was before another large convoy came in to anchor for the night.[21]

That afternoon, Lt. Colonel Protzman received orders to meet with Brigadier Generals Frederick Blesse and James Simmons who wanted to discuss future operations with him. Simmons (Chief Epidemiologist of the U.S. Army had just flown in from Washington, D.C.) on an inspection tour in North Africa. Blesse assured Protzman that *Acadia* would be part of the Salerno invasion fleet and wanted to make certain that she would be fully prepared to take care of a large number of wounded.[22]

The following day, *Acadia* embarked approximately 100 nurses for transport to Bizerte, Tunisia, who came aboard from a tank landing ship in fatigues (uniforms) and with field packs. Protzman described their composition and recent activities:

> They belonged to the 95th Evacuation and 16th Evacuation Hospitals, with another 2 girls from the 2d Auxiliary Surgical Group. They told us they had lived in the field for the past four months, housed in pup tents, eating in mess lines, and trying to stay in condition. Theirs was the first decent meal served at a table. They all had a nice tan and were ready for any emergency. They were an excellent object lesson for my own girls and made them feel that a Hospital Ship wasn't such a bad place after all, even though shore leaves are few and far between.[23]

Acadia departed Oran on 3 September for Bizerte. Protzman noted on the 4th, "The weather is fair and the girls from the Evacuation Hospitals are having the time of their lives. This is the first time they have slept in beds since landing in North Africa. My boys put on a show for them in the evening." The hospital ship arrived at Bizerte in late afternoon on 5 September, "docked at the silo pier and the Nurses marched off with their packs, were entrucked and whisked away." Protzman also commented on congestion/conditions at the port:

> Several ships of our old convoy are here in the harbour and we are anchored near the [troopship] S/S *Slooterdijk*... Three General Hospitals were landed here last night and will be set up as rapidly as possible. Tremendous preparations are under way for this attack against the mainland [Italy]. While we are at the pier, a steady flow of loaded LSTs is passing into the harbour for a trip to "where"? The pier alongside the dock is stacked with high explosives and artillery shells... If anything should drop on them during the coming night, we will cease to function! That's for sure.[24]

On the morning of 6 September, patients arrived on the pier at 0830 for loading aboard *Acadia*. Her medical staff was prepared having gotten up at 0530. After embarking both patients and prisoners of war,

Acadia left her berth and stood out of the harbour on transit to Oran. Going out the channel, she hit an unseen obstruction, and healed over. Believing the hospital ship had suffered harm to her hull or running gear, Protzman observed, "We need to go into dry dock and find how much damage has been done this time." (In spite of his remarks the ship continued on her way, but was docked later as expected.)[25]

On 8 September, *Acadia* was outside the breakwater at Oran, waiting for a large convoy to enter port before her. There were rumors circulating that Italy was going to capitulate before the invasion of Sicily. *Acadia* had on board many Italian prisoners of war—including general and other senior Army officers—who, it appeared, would soon be returned to Italy. Protzman explained:

> Along with our wounded, we had a great many prisoners of war, all Italians. There were 11 Generals, 27 Colonels, 19 Lieutenant Colonels, and 1,400 Enlisted Men. While we were on board, a few of the Generals complained that the shower water wasn't warm enough. After 2000 we received confirmation that Italy had surrendered and that all Italian PWs are being unloaded to be held here, and then sent back to their country. As there are still a great many German divisions in Italy, the going may be pretty tough before the place is completely secured and occupied by the Allied Forces.[26]

Protzman did not record in his diary when *Acadia*'s patients and prisoners of war were debarked. A week passed before the hospital ship was finally dry docked at Oran on 16 September. Protzman described the damage found and repairs made:

> We finally went into the floating dry dock at 1000 and by late afternoon our ship was high and dry. The port propeller had only a few pieces chipped from it when we hit a wreck at Ferryville, but the starboard one has one blade folded like an accordion, and two others badly bent. Only one blade is intact. After a long talk with the French master mechanic, he is confident that the blades can be straightened, but repairs will take several days. We further have a large dent on the starboard side of our ship where we hit that wreck at Bizerte and smashed into an eighty-foot-long keel. The *Acadia* has taken a terrific beating and some day we will probably have to return home for definitive repairs.[27]

Repairs progressed slower than desired, because French workers refused to work at night. Protzman appealed to the supervisor, citing the importance of *Acadia* returning to duty, and also give him some personal gifts, to expedite the work:

> I tried explaining that ours was a Hospital Ship, needed badly at the front, and with the aid of a good pair of army shoes, some soap, and some tobacco, he agreed to work with his men till midnight each night to finish repairing the propellers.[28]

On 18 September, Protzman described in diary entries that USAHS *Shamrock* had arrived at Mers-el-Kebir that day; that HMHS *Newfoundland* has been sunk, and HMHS *Lenster* damaged and abandoned; and that over one hundred Arab workers were employed scraping barnacles off *Acadia*'s hull in preparation for painting it. His entry regarding the British hospital ships and Arab workers follows:

> HMHS *Newfoundland* was hit and sunk off the coast of Italy four days ago, but luckily all on board were saved. [As detailed in the next chapter, all the doctors, six British nurses, and some of the crew were killed.] The [American] Nurses we brought to Bizerte were on that boat but none of them were hurt. Another British Hospital Ship, HMHS *Lenster* that laid in Lake Bizerte with us for so long was also hit and had to be abandoned. There are over a hundred Arabs scraping the barnacles from our ship's bottom so that it can be painted before we sail. Many of them fall asleep during work; as it is the Holy month of Ramadan, no food is allowed during day, and the men feel weak and are not very much motivated to work hard.[29]

At 1945 on 21 September, *Acadia* finally left dry dock with all repairs completed and a new coat of paint on her bottom. Boasting straight propeller blades once again, she could now, when desired or necessary, make best speed once again.[30]

12

British Hospital Ship *Newfoundland*

13 September D+4

0255 Enemy aircraft bombing shipping near port of Salerno.

*0401 Hospital ship LEINSTER bombed; position 40° 15' North, 14° 17'
East, at 0313.*

*0454 Hospital ship ST. ANDREW position 40° 13' 30" North, 14° 20'
East attacked. Hospital ship NEWFOUNDLAND on fire. ST.
ANDREW proceeding to NEWFOUNDLAND's assistance.*

*0816 It was reported that British hospital ship NEWFOUNDLAND had
been bombed and abandoned about 0500 in position 40° 14' North,
13° 20' East. Three other hospital ships and two destroyers standing by.*

—The Allied Invasion of Salerno, Italy (Operation AVALANCHE)
commenced on 9 September 1943 (termed D-Day). Four days
later, on the 13th, the British hospital ship *Newfoundland* was
bombed by enemy aircraft and set aflame. The unsalvable
ship was sunk by friendly naval gunfire on 15 September.

Photo 12-1

British hospital ship HMHS *Newfoundland*, in Casablanca Harbour, French
Morocco, on 14 November 1942, following the Allied invasion of North
Africa, and occupation of that port. A French tug is in the foreground.
National Archives photograph #80-G-30658

HMHS *Newfoundland* was the fifth British hospital ship lost to German aircraft attacks in World War II. She was built in 1925 by Vickers Ltd. Shipyard on Barrow Island in northwest England, for service as a British Royal Mail Ship. In this role, the 406-foot RMS *Newfoundland* worked a regular transatlantic mail route between Liverpool and Boston via St John's, Newfoundland, and Halifax, Nova Scotia, until requisitioned in 1940 for war service as a hospital ship.[2]

In the early years of the war, HMHS *Newfoundland* was employed sailing back and forth across the Atlantic between the UK and Canada carrying the wounded to Canada and bringing back rehabilitated troops. Under Captain John Eric Wilson OBE, she usually voyaged between Liverpool and Halifax, Nova Scotia.[3]

INVASION OF SALERNO, ITALY

In September 1943, when the Allies invaded Italy, HMHS *Newfoundland* was assigned as a hospital ship to Gen. Bernard Montgomery's British Eighth Army. *Newfoundland* was one of two hospital ships sent to Salerno beaches on 12 September, where they were to deliver 103 embarked American nurses. U.S. Army nurse Vera Lee described events of that day in a letter:

> We tried to land in Italy all day Sunday the 12th but they were too busy fighting to worry about a hundred nurses on a hospital ship. Several bombs just missed us several times but we didn't really realize what it was all about. Evening came & we had to go out of the harbour because our ship was all lit up. We taxied around in the sea off shore about 30 miles all night—our ship & 4 other Hospital ships—at 5 a.m. we were awakened by a bomb falling very close to us—Some of the girls dressed then but most of us went back to sleep.[4]

The preceding evening, there had been a party on board held in honour of the embarked American Nursing Corps. *Newfoundland* had only two patients, so it had been quiet for all the nurses and doctors. Afterward, British Sisters Dorothy Mary Cole and Vera Schofield showed some of their American counterparts around the ship's operating theatre.[5]

Lee continued her account in the letter:

> At 5:10 we heard a plane & then that bad awful whistle a bomb makes & bang!—You'll never know of the thousand things that flashed through my mind those few seconds. I thought sure I was dying—could feel hot water falling on my face & body—Had heavy

boards on my chest that had fallen from the ceiling—I shut my eyes & thought it was the end—Then the next second I thought "What the hell, I'm not dead—get out of this place"... I couldn't see for the terrific smoke in our room—but was a mass of motion trying to find my coveralls which...I found on the floor—all soaked with water & black with dirt—put them on & found my shoes—grabbed my helmet & water canteen & grabbed on to someone's arm & followed the light that Claudine was holding. She couldn't hardly find where the door was because the wall had all been blown out.

When we got on the deck we all had to get on one side because the bomb had torn away the other side of the ship.... We loaded in a life boat—70 of us in one boat that had a capacity of 30. Were taken on another hospital ship & given tea & hot coffee. I felt a darn good cry coming on so some British fellow took the 4 of we girls to his room & we drank a bottle of Scotch. I got "stinko" drunk—cried & when I snapped out of it, I felt fine. All the bruises I got out of it was a scratch on my knee, a cut on my left foot and marks & scratches on my chest where debris fell from the roof.[6]

At around 0500 that morning, there came the sound of a single aircraft, and Captain Wilson, who was on the bridge, heard a bomb falling. It landed on the boat deck behind the bridge and exploded, causing a lot of damage and setting the ship on fire. Communications were lost and, more importantly, firefighting equipment was rendered inoperable.[7]

All of the doctors, six nurses, and a number of the ship's officers were killed. The surviving British nurses and all the American ones went straight to their stations in the smoke and flames and waited to be told what to do. There was another explosion, setting the fuel oil tanks on fire, and the order was given to abandon ship. The survivors took to the lifeboats. The 2nd Officer, with a broken leg, a broken arm and splinter wounds, was in one of the boats full of nurses.[8]

UNSUCCESSFUL EFFORTS TO SAVE THE SHIP

At 0454/D plus 4 [13 September] the British hospital ship NEWFOUNDLAND was bombed and set [on] fire by enemy aircraft in position Latitude 40° 13' 30" North, Longitude 14° 20' East. Three other hospital ships and two destroyers stood by to assist the NEWFOUNDLAND which was abandoned. At 1150 Comdesron 7 [Commander Destroyer Squadron 7] reported that the hull of the ship was intact and that the fire was now confined amidships. About 12 hours or better would be required to extinguish the fire.

> *Although strenuous efforts were made to save the hull, it became apparent that*
> *effective salvage would be impracticable and she was ordered to be sunk.*
>
> —Commander United States Eighth Fleet,
> The Italian Campaign, 11 January 1945.

Captain Wilson and seventeen volunteers stayed on board to fight the fire, assisted by USS *Mayo* (DD-422) and USS *Plunkett* (DD-431) which came alongside the *Newfoundland*. The two destroyers had been returning to the assault area at Salerno Bay—after escorting the cruisers USS *Philadelphia* (CL-41) and USS *Boise* (CL-47) to latitude 39° north— when an explosion was sighted at 0515 bearing 300°T. Ten minutes later, the ships changed course to 060°T. At 0530, a ship (which turned out to be HMHS *Newfoundland*) was sighted, apparently on fire, bearing 290°T.[9]

At 0612, *Mayo* was directed to investigate the fire; *Plunkett* was similarly directed at 0627 to proceed to the fire location. *Plunkett* went alongside the *Newfoundland* at 0730, and started to fight the fire. *Mayo* was already alongside, and survivors had been removed to other hospital ships in the area. At 0900, the master of the *Newfoundland*, Captain Wilson, came on board *Plunkett*.[10]

At 0930, *Plunkett* stood away from *Newfoundland* as she was gaining list (heeling further over) at that time, but came back alongside her ten minutes later, and sent on board a firefighting party. The fire was confined to the centre portion of the ship, and *Mayo* was ordered at 1045 to leave her side and take up patrol.[11]

At 1125, the fleet tug USS *Moreno* (AT-87) arrived in the area with Commodore William A. Sullivan aboard who took charge of the fire fighting. Sullivan, Chief of Naval Salvage, had been detailed to duty with the U.S. Eighth Fleet in the Mediterranean. *Plunkett*'s firefighting party returned on board the destroyer, except for two men who were detailed to remain on board and handle lines for the *Moreno*.[12]

The fleet tug moored at 1134 with her port side to the port side of the hospital ship. Such an arrangement, with the bows of two ships alongside one another pointing in opposite directions, is termed a "Chinese moor." The tug's firefighting and salvage repair parties then began trying to extinguish fires aboard the ponderous vessel, which was listing to starboard. In the early afternoon as firefighting efforts continued, nineteen officer and enlisted survivors of the stricken vessel came aboard the tug.[13]

Twenty-four hours later, after all the fires were finally extinguished, *Moreno*'s salvage repair party began trying to correct the list of the *Newfoundland* through use of the tug's salvage pumps. As salvage efforts were not succeeding, they were ceased in the early evening following receipt of orders to stop work and await instructions regarding the disposal of the hospital ship, that *Newfoundland* would be abandoned.[14]

At 2115 on 14 September, *Plunkett* on orders fired on and sank the hulk. *Moreno* then proceeded for the Gulf of Salerno. The tug anchored four miles offshore ten minutes past midnight and Commodore Sullivan departed her a half hour later. He had felt that *Newfoundland* could have been saved and, when the order came to sink her, sent a message recommending she be left alone until a salvage tug could take her under tow for a repair facility or at least move her into shallow water. After two hours as no answer was forthcoming, the destroyer sank the hospital ship. It was later learned that both Admirals Henry K. Hewitt (commander, U.S. Eighth Fleet) and Sir John Cunningham (commander-in-chief, Mediterranean Fleet) had approved the request, but neither message was received.[15]

Photo 12-2

L-R: Vice Adm. William A. Glassford, USN; Adm. Sir John Cunningham, RN, Commander-in-Chief, Allied Naval forces Mediterranean; and Adm. Henry K. Hewitt, USN, on 4 May 1945. On this date, Glassford relieved Hewitt as commander of U.S. Naval Forces, Northwest African Waters.
National Archives photograph #80-G-316733

LEST THEY BE FORGOTTEN

Six of the British nurses on board HMHS *Newfoundland* were lost as a result of the German aircraft attack. Four of those killed were members of the Queen Alexandra's Imperial Military Nursing Service (QAIMNS), and the remaining two of the Territorial Army Nursing Service (TANS):

- Matron Agnes McInnes, QAIMNS
- Sister Una Cameron, TANS
- Sister Mary Lea, TANS
- Sister Dorothy Mary Cole, QAIMNS
- Sister Phyllis Gibson, QAIMNS
- Sister Margaret Annie O'Loughlin, QAIMNS[16]

British vessels rescued the remaining British and American nurses and evacuated them to Bizerte, Tunisia. Four U.S. Army nurses suffered minor wounds for which they later received the Purple Heart. The others boarded another ship and returned to Salerno ten days later.[17]

USAHS *Acadia* and *Shamrock*

Photo 13-1

U.S. Army Hospital Ship USAHS *Shamrock* in port, circa 1943-1946.
Naval History and Heritage Command photograph #NH 103128

The elderly 429-foot hospital ship *Shamrock* was built in 1906 by William Cramp & Sons Ship & Engine Building Co., Philadelphia, Pennsylvania. Originally named *Havanna*, she operated commercially for the New York & Cuba Mail Steamship Co. until 1917. Acquired by the War Department, she was subsequently transferred to the Navy, and then outfitted at the New York Navy Yard in Brooklyn, New York, as the hospital ship USS *Comfort*, commissioning on 18 March 1918.[1]

After serving from 24 July to 5 October 1918 as a floating hospital at New York, *Comfort* was employed returning wounded servicemen from Europe. Between 21 October 1918 and 13 March 1919, she brought home 1,183 men from France, Britain, and the Azores. Following this service, she was decommissioned on 5 August 1921, and subsequently sold to General Metal Supply Co., Oakland, California, for scrapping—however, this company ultimately resold her intact.[2]

In 1928, *Comfort* was renamed SS *Havana*, subsequently *Yucatan*, and later (in 1941 following purchase in 1941 by Agwilines, Inc.) *Agwileon*. Little is known about her service in this period. She was bareboat chartered by the Army in November 1942 and underwent conversion to a troopship by Atlantic Basin Iron Works, Brooklyn, New York, between November 1942 and April 1943. In late April she left for Oran and Gibraltar. Following this service and her return to the United States, she was converted to a hospital ship by Atlantic Basin Iron Works, and was renamed *Shamrock* in accordance with a recently adopted policy of the Surgeon General's Office to have hospital ships named after flowers.[3]

NAMING OF U.S. ARMY HOSPITAL SHIPS

First Lieutenant (Nurse Corps) Aleda E. Lutz, United States Army Air Forces, was awarded the Distinguished Flying Cross (Posthumously) for extraordinary achievement while participating in aerial flight while serving as a Flight Nurse with the 802d Medical Air Evacuation Squadron during World War II

Prior to 1943, there was no U.S. Army policy regarding the naming of hospital ships. Ships generally retained their original, or most recent name while purchased or charted for the service of the Army. Such was the case with the previously mentioned *Acadia* and *Seminole* which were employed in the summer of 1943. Also, it was considered there was no need to rename them, since both ships were well known and new names might confuse the enemy who in contrast to other ships needed to be made aware of the identification of hospital ships. However, as the Army selected additional vessels for conversion to hospital ships, it became advisable to adopt a definite naming policy.[4]

On 1 July 1943, the Office of the Surgeon General recommended to the Adjutant General that:

In order not to trespass on the series of names given to various types of naval vessels, it is recommended that Army hospital ships be named for flowers. Flowers indicate a quality of mercy and from many are obtained curative drugs used in the treatment of the sick.[5]

In a subsequent, related action, the Chief of the Water Division, proposed to the Chief of Transportation on 22 July that the names *Shamrock*, *Larkspur*, *Marigold*, *Wisteria*, and *Thistle* be used for the next five hospital ships. Ultimately, this convention was followed for the naming of six hospital ships: when *Dogwood* was added.[6]

The choosing of names for Army hospital ships was more important than it might appear, since each vessel had to be registered officially as to name and characteristics to be readily recognizable by the enemy. In addition, a serial number was also required which had to be prominently displayed on each vessel's hull. Once a name had been registered, any change had to be communicated to the enemy through the State Department and the Swiss Government. Moreover, as previously noted, there were some vessels so well known in the shipping world, that renaming them was considered inadvisable. Thus, a number of hospital ships would serve under their old names.[7]

In March 1944, following deliberation on the matter by the Office of the Surgeon General, a decision was made to name the rest of the Army hospital ships after deceased Army doctors and Army nurses who had served with distinction. The first person so honoured was Emily H. M. Weder, an Army nurse since 1918, who died on 10 September 1943 at the Walter Reed General Hospital, Washington, D.C.[8]

During her early military service in 1919, the Army had sent Weder to serve in Siberia with fifteen other nurses supporting the American force deployed to contain Communist revolutionaries. After six months of difficult duty in the Siberian winter, the Army had then ordered Weder to the Philippines for three years. In 1943, after many more years of distinguished service, Major Weder was preparing to leave her post as assistant chief nurse at Walter Reed General Hospital in Washington, D.C. for duty off the south of France to support the expected Allied invasion. At this time Welder was in charge of all operating rooms at Walter Reed. Sadly, before being deployed, she was diagnosed with cancer and died at the age of forty-nine. Maj. Emily H. M. Weder was buried with military honours at Arlington Cemetary, near a monument dedicated to women who died in the military during World War I.[9]

Nine Army doctors/nurses including Weber ultimately became the namesakes for hospital ships:

- Aleda E. Lutz: Lt./Flight nurse killed in a plane crash in southern France in November 1944, after having participated in 190 missions to evacuate wounded personnel by air. First military woman to receive the Distinguished Flying Cross, which she received posthumously.
- Blanche F. Sigman: 1st. Lt./first Army nurse to be killed in action on the Anzio beachhead
- Charles A. Stafford: Capt./Medical Doctor posthumously awarded the Silver Star for his actions during the evacuation of Java
- Emily H. M. Weder: Maj./Nurse served in World War I on a special detail with the AEF at Vladivostok, and subsequently for three years in the Philippines and had important assignments at Walter Reed and Letterman General Hospital before her death in 1943
- Ernestine Koranda: Lt./Flight Nurse killed in a plane crash in the Southwest Pacific on 19 December 1943
- Frances Y. Slanger: 2nd Lt./seventh nurse to lose her life in World War II, and the only American nurse to die due to enemy fire in the European theatre of World War II
- Jarrett M. Huddleston: Col./Surgeon in the Fifth Army, who was killed in action in Italy early in 1944
- John J. Meany: Maj./(Presumably a medical doctor or surgeon) killed in action in North Africa in March 1943
- Louis A. Milne: Col./Port Surgeon at the New York Port of embarkation from 1937 to the time of his death in 1943

U.S. Army Hospital Ships in World War II (24 total)

Retained Names	Flower Names	Tribute Names
USAHS *Acadia*	USAHS *Dogwood*	USAHS *Aleda E. Lutz*
USAHS *Algonquin*	USAHS *Larkspur*	USAHS *Blanche F. Sigman*
USAHS *Chateau Thierry*	USAHS *Marigold*	USAHS *Charles A. Stafford*
USAHS *Ernest Hinds*	USAHS *Shamrock*	USAHS *Emily H. M. Weder*
USAHS *John L. Clem*	USAHS *Thistle*	USAHS *Ernestine Koranda*
USAHS *Republic*	USAHS *Wisteria*	USAHS *Frances Y. Slanger*
USAHS *Seminole*		USAHS *Jarrett M. Huddleston*
USAHS *St. Mihiel*		USAHS *John J. Meany*
USAHS *St. Olaf*		USAHS *Louis A. Milne*

ALLIED MOVEMENT NORTHWARD IN ITALY

Returning to the subject at hand, following the battle for the Salerno beaches, the subsequent Allied campaign to take Naples and Foggia to the north proceeded slowly and methodically. Montgomery's British Eighth Army, sweeping around the right flank of Clark's U.S. Fifth Army, crossed the Ofanto River in southern Italy near the Adriatic coast and, on 27 September, drove through Foggia in southeastern Italy into the hills that commanded the area where Italian Royal Air Force airfields were located. Elements of the Fifth Army reached Benevento on 2 October.[10]

Progress was slow because the terrain was mountainous, and the Germans, while retreating northward, employed skilled delaying tactics. Rear guard detachments on selected hillsides using mobile artillery, forced Allied troops to make slow enveloping movements; bridges were blown by the enemy to slow advancement; deadly land mines were plentiful; and towns were filled with rubble to second-story level.[11]

EMBARKING PATIENTS OFF SALERNO BEACHES

1330 [24 September 1943] and we are off the Isle of Capri. We can hear and even feel heavy detonations coming from the direction of Naples, Italy, where the Germans are destroying the docks and the water front. There is heavy smoke all along the coast and the sea is filled with debris. The coastline in front of us is dotted with destroyers, and one of them just passed us towing a wrecked LST. Allied bombers and fighters are overhead. There seems to be continuous firing from the hills and it looks as though we might spend an interesting evening. USAHS Shamrock *is anchored in the neighborhood and taking on patients bound for Bizerte, Tunisia. She will leave the next morning.*

It is 1900 and we have been ordered to sea (opposite the little town of Agropoli) [at the southern end of the Gulf of Salerno] till morning as it isn't safe for us to remain here during the night. We will come back Saturday morning and begin loading wounded and then start back for Oran, Algeria. As we left, ground forces were putting on some smoke screens over the shore and the fleet for night protection.

—Lt. Colonel Thomas B. Protzman diary entry regarding the locations/ activities of USAHS *Acadia* and *Shamrock* on 24 September 1943.[12]

Although this chapter endeavours to honour the hospital ship USAHS *Shamrock*, information about her activities is scarce; the author found little beyond other units' brief references to her in their war diary entries. Accordingly, material about the backdrop against which she operated in Italy, is largely gleaned from diary entries by Lt. Colonel Thomas B. Protzman, about the activities of *Acadia*, and efforts by her medical staff to care for wounded personnel during their transport to shore hospitals more distant from the front lines.

On the morning of 23 September 1943, *Shamrock* stood into the Gulf of Salerno to load wounded. Two days later, commander, U.S. Eighth Fleet sailed her (directed her to proceed) for Bizerte, Tunisia, via Palermo, Sicily.[13]

On the night of 24-25 September, *Acadia* circled outside the harbour of Paestum, Italy, marking time near Red Beach #2. Around 0800 on the 25th, she moved in closer to the beachhead, and Protzman went ashore in an assault landing craft, accompanied by Brig. Gen. Frederick A. Blesse and Colonel Reeder (who were embarked aboard the hospital ship) and some of Protzman's officers to see about patient embarkation plans. After beaching, the group made a quick run to the frontlines only a few miles away in a borrowed jeep, before returning aboard ship. Protzman noted the following about this reconnaissance:

> We couldn't see much of the hills and fortunately escaped some of the enemy heavy shells that were bouncing around. This is the first actual beachhead that we visited, as the places on Sicily were always in or near some city. The area is still heavily mined and we were to walk only within designated paths and roads, and only there, with caution.[14]

Map 13-1

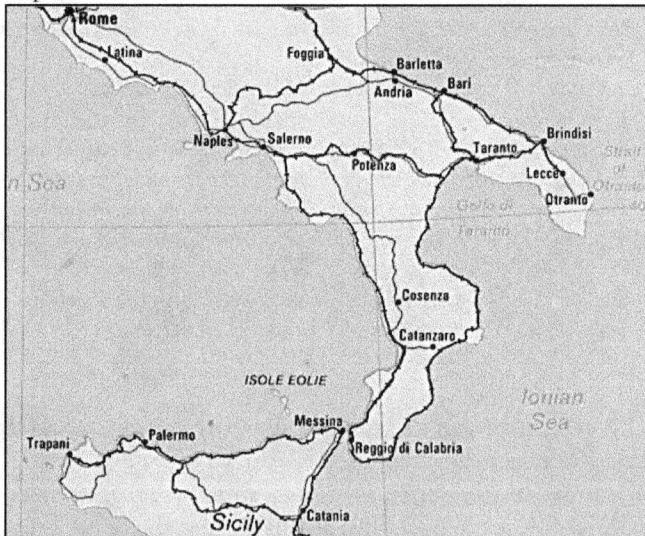

The Gulf of Salerno lies off Salerno (east-southeast of Naples); and the Isle (Island) of Capri a little to the northwest, just off the peninsula forming the northern boundary of the Gulf of Salerno and southern boundary of the Bay of Naples

That afternoon, 25 September, Protzman took 60 enlisted men and 18 nurses ashore to assist with embarkation of the wounded in the LCTs (tank landing craft) that would bring them to *Acadia*. When he returned aboard *Acadia*, the survivors of the minesweeper USS *Skill* (AM-115) torpedoed by *U-593* (Kptlt. Gerd Kelbling) just eight miles from the hospital ship, were being brought alongside in USS *Speed* (AM-116).[15]

Skill had been on anti-submarine patrol station "Nan" between Point Licosa and Isle of Capri in the Gulf of Salerno. After arriving at her patrol line, at 1140, a terrific explosion occurred without warning caused by a torpedo hit below the waterline between frames 35 and 45. The warhead detonation caused an explosion in her forward magazine, breaking the ship in two at about frame 43. The bow section capsized, and fire broke out in the forward part of the after section.[16]

Speed picked up thirty-five survivors, and proceeded at 1330 to the *Acadia*, anchored in the southern area of Salerno Bay. Sadly, two of the *Skill* crewmen died en route and another later also perished. Protzman described the actions of his medical personnel to provide care for the survivors (amid naval shore bombardment and fighting ashore), and subsequent recognition of their efforts:

> They blinked for help [flashing light message from *Speed*], and we despatched three shock teams of Enlisted Men and Nurses by small boat to them with medical supplies and plasma. We worked all through the night to help the survivors, 33 of them, all burned and terribly torn by the explosion and concussion. Our ship remained in the beach area all night and we had a much closer view of the artillery and counter battery fire going on in the hills. It was midnight and up on deck again after a slight rest to watch British battleships and cruisers throwing heavy shells into Naples and surroundings.

> We finished operating [on] the last man at 0430 this morning [26 September]. We received a Commendation from the Admiral of the Fleet for saving the majority of the minesweeper's survivors.[17]

At noon on the 26th a landing barge with 112 wounded pulled alongside the *Acadia*. A heavy sea running, and they were brought on board in about fifty minutes with help from Chaplain William V. Barney's loading teams. Fire and smoke drifted over Naples, and it appeared that there was a heavy and long hard fight going on there. All day long patients trickled onto the ship, brought to her in all sorts of conveyances, and in all sorts of condition.[18]

The following morning, 27 September, *Acadia* relocated off Blue Beach to pick up more wounded. Several were brought by small craft. The waves were so high, loading was extremely difficult but it was accomplished efficiently with all the patients brought safely aboard. At 1010, word came from the shore that it was impossible to load patients on the beach that day; consequently at 1515, *Acadia* moved off the North Attack Beach, just outside of Salerno harbour. As she anchored, two German planes were shot down overhead.[19]

Protzman noted, "There's so much action here and everything happens so fast that it is difficult to follow," and described patient care/combat action that day:

> 2100; we just finished amputating a leg at the hip. Shrapnel wounds tear up the blood vessels to pieces, gangrene sets in, and the limb has to be removed well above the injury, to save the patient's life. The North Beach Invasion Fleet is anchored a mile from us and a heavy shore battery is sending shells into the nearby hills. Somewhere across the mountains, a German counter battery has been trying to locate our ships and our shore battery. Many of their shells have landed disconcertingly close to the *Acadia*, rocking the boat [hospital ship] and scaring the hell out of our patients and us. Two British Hospitals [ships] are also in the area. It's understandable to see why we landed here and captured the coast, there are at least four airfields that we can see from the ship and planes of all types are constantly in the air.[20]

Surgical staff aboard *Acadia* finished their work early the next morning (around 0130 on 28 September), and afterwards with the exception of one shell that landed close by at 0330, slept undisturbed. Protzman described in diary entries: losing a patient who had suffered internal injuries as a result of the force of an exploding munition; receiving wounded British troops on board; necessity to move the ship because of a mine threat; reasons for delays in receiving additional patients; and development of a gale and associated rough seas:

> A second blast injury died today. These patients were literally blown to pieces inside, yet there isn't any mark on the outside of their bodies…. All the patients we took on today came from the British Forces that are carrying on the attack in the hills, and they're doing a splendid job. Colonel Sir Terence Edmond Patrick Falkiner of the Coldstream Guards is one of our patients. He told us of an amusing incident while staying on the Island of Pantelleria [in the Strait of Sicily] with American troops. While mostly subsisting on American food and field rations, he came upon a plug of chewing tobacco

with a wrapper describing it as delicious and dipped in honey. The Officer told his staff; "The Americans are our Allies, and it is only right that we learn to eat their food." They did sample the tobacco, and later vomited the whole meal as a consequence.

At 1400 a British minesweeper ordered us to move at once as we were drifting over a German radio-controlled mine.

At present we only have 250 patients on board as the rough weather has made it impossible to get the wounded onto the landing craft. Another delay was caused by the moving of some British armoured elements over the narrow mountain roads blocking all return of the ambulances evacuating the wounded. Naples will be bypassed as the enemy seems to have destroyed the city and sunk all vessels in the harbour.

At 2200 we finished operating for the day. A high gale developed, necessitating the dropping of another anchor, as we were drifting away with the high winds.[21]

On the morning of 29 September, *Acadia* moved from the North Attack Beach back to Red Beach #2. Protzman did not understand the reason for her relocation because the seas were too high to make it possible for any wounded to be loaded in boats on the shore for transfer to the hospital ship. He described the conditions and challenges that existed that day off the Salerno beaches:

The storm was far greater than we imagined as a great deal of the lighter vessels have been washed ashore, and it will be some time before they will be serviceable again. Our ship is so close to shore that the mountains interfere with our radio communications and we haven't had any news until this morning. Two more enemy aircraft have been shot down. The ocean around us is filled with small landing craft loaded with troops and cargo that will land as soon as the weather subsides. We had to take off a boy's arm this afternoon; it depresses me every time this becomes necessary, but this isn't nearly so bad as the young lad with both eyes gone.[22]

Month's end found *Acadia* still at Red Beach #2 bordering Green Beach, and *Shamrock* back in the area from a trip to Bizerte. *Shamrock*'s commanding officer, Colonel Davis came aboard *Acadia* accompanied by his supply officer and American Red Cross worker. He brought some extra blood plasma, *Acadia*'s supply being low as a result of the large number of shock and hemorrhage cases she had treated in the past few days. Protzman's diary entries for 30 September include:

Smaller craft are bringing in wounded now. Naples is still in enemy hands, it seems that the hotels and large buildings have been destroyed by the Germans and our own fire, the water viaducts have been blown up, and the city is becoming disease ridden.

Our greatest diversion today consisted in watching the antiaircraft batteries from the warships try to bring down the high-flying German observation planes that keep crossing over our area. So far no planes have been hit, nevertheless we as innocent bystanders have to watch out for falling flak fragments.

1630 and we have taken on 204 more wounded.[23]

That night 30 September, loaded with available patients, Protzman expected but did not receive sailing orders for *Acadia* to proceed to Oran to debark her British patients. The following morning, with the hospital ship still standing by off Red Beach #2 awaiting direction from commander, Eighth Fleet, Protzman went ashore in an Army DUKW (amphibious truck). After first visiting U.S. Fifth Army headquarters, he went directly to the destroyer aboard which the admiral was embarked. There, he learned that *Acadia* would be re-routed to Bizerte to load additional patients, to full capacity and then sail for Algiers.[24]

Photo 13-2

U.S. Army DUKW amphibious truck bringing supplies
ashore on a Normandy beach, 11 June 1944.
National Archives photograph #80-G-252737

Protzman noted in a diary entry on 1 October, his concerns about entering port at Algiers, and pride in the performance of his Hospital Unit personnel on board *Acadia*:

I certainly hope we'll get in and out of this place without damages, as the least little bump on the propellers now, will break them, and we'll get stuck here. We have a bad case of dog bite aboard and we have no anti-rabies vaccine, and there may be none in Algiers. I cannot help but mention the great performance of my outfit; in this last trip my men have saved a great many lives and besides doing the actual work, most of them gave their own blood for transfusions in their efforts to save lives.[25]

PORT OF NAPLES OPENED

Hundreds of vessels and craft such as floating sheerlegs, cranes, lighters, barges, tugs, small tankers, trawlers, corvettes, sloops, sailing ships, landing craft and destroyers were available to him [retreating German commander] for employment with merchant ships in systematically blocking the major and minor ports.

The wreckage of these vessels obviously had been supervised by an expert with a knowledge of salvage, for each vessel had been methodically destroyed internally in the way of bulkheads, and then so badly "blown" as to preclude use of either pumping or compressed air.

—Commander Eighth Fleet, Vice Adm. Henry K. Hewitt, USN, describing the devastation wrought by German forces to the port and harbour facilities at Naples, Italy, prior to their withdrawal from the city.[26]

The U.S. Army's 82nd Airborne Division, reinforced by the Ranger Brigade, captured the city of Naples on 1 October following which efforts immediately began to open the port. The approach channels and harbour were swept and the work necessary to make the port operational was begun. The Germans had thoroughly demolished the harbour craft, cranes, and port operation equipment, and sunk quantities of vessels to prevent Allied use of the port. Making matters worse, over each sunken vessel, retreating forces had:

Invariably sunk a few lighters, a dock crane, an occasional locomotive or a string of trucks, upon which wagon-loads of ammunition, oxygen bottles, and small arms had been dumped in a haphazard manner. This super-tangle of obstructions imposed extensive diving operations even after the main objective had been reached, and many wrecks had been sunk in such a manner as to be invisible from the air.[27]

PROGRESS MADE, BUT...

[It was] the most daring amphibious operation we have yet launched or which I think has ever been launched on a similar scale. We must...consider this episode—the landing on the beaches of Salerno—as an important and pregnant victory, one deserving a definite place in the records....

—Excerpts from British Prime Minister Winston Churchill's summarisation of Operation AVALANCHE on 21 September 1943, in which he also highlighted that intervention of strong naval gunfire support forces and air forces had played a large role in the recovery of an unfortunate military situation.[28]

As Allied forces occupied the city of Naples, the Germans were withdrawing to the Volturno River and trying to establish a defensive line across the Italian peninsula. With enemy forces retiring northward, and the Allies having established two armies on the Italian mainland—the U.S. Fifth and British Eighth—the prospects for advancing rapidly to Rome appeared to be good. The Allies did not yet appreciate the extent to which the Germans could use the Italian winter weather, the Italian terrain, and the mettle of their own outnumbered troops to deny the Allies and the Italians quick entry into the capital city.[29]

Crossing the Strait of Messina had been easy, but securing a beachhead at Salerno proved to be more difficult. Though much loss of life was suffered by both sides, neither achieved its objectives. The Germans had wanted to drive the Allies off the beaches and had failed. The Allies were ashore and their position in Italy would never again be seriously threatened. But their hopes of using the surrender of Italy to make quick gains advancing northward up the peninsula toward Rome had not been realised.[30]

These setbacks would be miniscule, though, in comparison to what lay ahead. No one could foresee the bitter ground-fighting that would take place at the Volturno and Sangro Rivers, at the small Adriatic Sea town of Ortona, on the approaches to the Liri valley, along the Rapido and Garigliano Rivers, in the shadow of Cassino, and on the Anzio beachhead. No one could anticipate the expenditure of men and materiel that would be necessary for the Allies to take Rome, least of all the Italians, who on 13 October 1943 would declare war on Germany.[31]

Lady Nelson's Support of the Allied African / Italian Campaigns

In WWII when Britain indicated she needed help with hospital ship capacity, Canada converted two former merchant vessels—*Lady Nelson* and *Letitia*. *Lady Nelson* (hull number 46) was a former ocean liner which had been torpedoed by the German submarine *U-161* while alongside the pier at Castries, St. Lucia, but later raised and repaired. Ex-passenger ship *Letitia* (hull number 66) had previously served Britain in the war as an armed merchant cruiser, and later as a troop ship, before being taken over by the Canadian Government and converted into a hospital ship.

Photo 14-1

Hospital ship *Lady Nelson*.
Credit: Canada. Department of National Defence/
Library and Archives Canada copy – ID number 4232592

The first to be commissioned as a hospital ship under the Hague Convention was the former 7,988-ton Canadian National Steamships,

West Indies cruise liner RMS *Lady Nelson*. Built in 1928, she was one of five sister ships named after the wives of British admirals with a connection to the West Indies. These vessels were affectionately known as the Lady Boats and provided an efficient cargo service and romantic cruises for many years, plying the old Canada-West Indies trade routes from the days of sail. Propelled by steam turbines coupled to twin screws, the 420-foot *Lady Nelson* could maintain a cruising speed of 14 knots. In addition to carrying passengers, the Lady Boats delivered Canadian goods to the West Indies, returning with bananas and other fresh tropical fruit in their large, refrigerated holds.[1]

In early morning darkness on 22 March 1942, *U-161* (Kptlt. Albrecht Achilles) fired two torpedoes at 0449 into the small harbour of Port Castries, St. Lucia, from its entrance. (The German submarine then backed her way out to open sea and made her escape.) The first torpedo hit the *Lady Nelson* (master George W. Welch) which caught fire and sank by the stern in shallow waters. Three crew members, 15 passengers, and two gunners aboard *Lady Nelson* and seven dock workers were lost. The second torpedo struck the British merchant vessel *Umtata* which also sank by the stern. Her casualties were four crew members, four gunners and 33 passengers killed. Both vessels were later salvaged and repaired.[2]

On 16 April 1942, *Lady Nelson* was salvaged, and temporarily repaired, then left for Mobile, Alabama, on 11 May. During her conversion to Canada's first hospital ship, she was stripped of her stately cabins in order to make room for hospital wards and medical laboratories. Necessary work was completed in April 1943. Now fully equipped and possessing both an operating room and X-ray facilities, she was commissioned HMCHS *Lady Nelson* on 22 April 1943.[3]

LADY NELSON'S HOSPITAL SHIP SERVICE

Lady Nelson then began a lengthy tour of duty as a mercy ship. Her Royal Canadian Army Medical Corps (RCAMC) was comprised of 9 Medical Officers, 14 Nursing Sisters, and 60 other ranks. Spiritual care was provided by two Army Chaplains, Protestant and Roman Catholic. During this time her medical facilities were under the command and care of Lieutenant Colonel Alexander Harold Taylor RCAMC. Fitted with 515 beds to berth patients, she would complete thirty unscathed voyages by February 1946.[4]

In early 1943 British authorities pressed Canada to devote *Lady Nelson* to sailing exclusively between Britain and the Mediterranean. Following refusal of this suggestion, one followed to staff a former French hospital ship specifically to serve this area. Unable to comply,

Canada made a comprise and modified *Lady Nelson*'s sailing arrangement:

> ...but it was finally agreed that the *Lady Nelson* operating as a Canadian hospital ship on the Atlantic run between Canada and the United Kingdom would, on her eastbound trip, call at North Africa and evacuate casualties both British and Canadian to England.[5]

Between 3 July and 12 December 1943, *Lady Nelson* made eight voyages from Halifax, Nova Scotia, into the Mediterranean. After loading soldiers at Algiers, Algeria, and/or Bizerte, Tunisia, wounded in the Allied Invasion of North Africa, she transported them to England or Canada. During the ensuing Italian Campaign, *Lady Nelson* twice came to Naples, Italy, to evacuate Canadian wounded from this theatre of operations. Her first mission to Naples commenced in Halifax on 14 January 1944. Her return voyage began at Naples on 12 March and ended upon her arrival back in Halifax on 12 April 1944. (It's unknown to the writer whether she received interim tasking, such as transporting patients to England before returning to the Mediterranean.)[6]

Photo 14-2

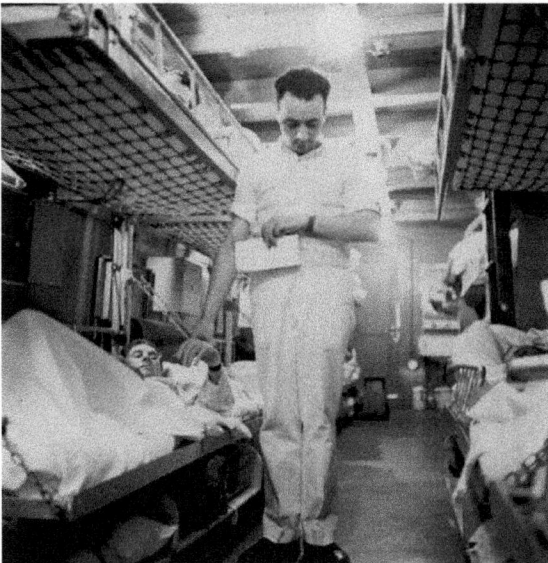

Hospital orderly E. Hartz checks the condition of Private F. E. Arden in a ward aboard the hospital ship *Lady Nelson*. Naples Italy, 29 January 1944. Credit: Lieut. Frederick G.D. Whitcombe/Canada Dept of National Defence/ Library and Archives Canada/PA – 163662 - ID number 3227114

Before departing this chapter devoted to the contributions of HMCHS *Lady Nelson* to the Allied Italian Campaign, it is appropriate to acknowledge the contributions of Canada's Nursing Sisters. Some of the sisters served on board the hospital ships *Lady Nelson*, and *Letitia* which is taken up later in the book.

CANADIAN NURSING SISTERS

> *In whatever conflict Canadians have been called on to bear arms, in the last hundred years, the medical services of Canada have earned a high reputation for the skill and devotion with which they played their special part. It is a reputation that has not suffered as they carried out their continuous function in time of peace. Canadian nursing sisters are justifiably proud to have borne their share alike with officers and men in the great contribution made by the medical services. Of these dedicated women it may be said "They served equally."*

—Closing paragraph of Col. Gerald W. L. Nicholson's
book, *Canada's Nursing Sisters.*

Photo 14-3

Nursing Sister with wounded Canadian soldier.
Cover of *Star Weekly*, 23 August 1941

Following Germany's invasion of Poland in September 1939, Canada again found herself thrust into a world conflict and, as in the preceding Great War, Nursing Sisters answered the call of duty. During WWII, they served at home and abroad in England, France, Belgium, the Netherlands, Italy, Algeria, Hong Kong and South Africa caring for Canadian troops, local civilians and prisoners of war. Concurrent with a dramatic increase in the size of Canada's military, the nursing service was expanded beyond the Canadian Army Nursing Service of the preceding war, to all three branches: the Canadian Army, the Royal Canadian Navy and the Royal Canadian Air Force.[7]

The nurses in each branch had their own distinctive uniform and working dress. All wore the Nursing Sisters' white veil, and were respectfully addressed as "Sister" or "Ma'am" because they were all commissioned officers. By war's end 4,480 Nursing Sisters had joined the services: including: 3,656 with the Royal Canadian Army Medical Corps, 481 with the Royal Canadian Air Force Medical Branch, and 343 with the Royal Canadian Naval Medical Service.[8]

The remainder of this section is devoted to the Army sisters who, after training in Canada, were the first to go overseas—to the United Kingdom where they supported combat units which had preceded them. Accompanying soldiers sent overseas, the sisters travelled by ship in large convoys, running the perilous gauntlet of German submarine action in the North Atlantic. Upon arrival in England, they were assigned duties at the Royal Canadian Army Medical Corps' hospitals at Taplow (located in Buckinghamshire on the left bank of the River Thames), and Bramshott and Basingstoke (both in Hampshire in south-central England).[9]

After three years in England, Nursing Sisters were sent into action in the Mediterranean. Upon the arrival of No. 1 Canadian General Hospital in Sicily, they wore battledress similar to all other Canadian soldiers. Almost all hospital units deployed to the continent were initially set up under canvas, and later were relocated into abandoned or bombed-out buildings.[10]

The second General Hospital was deployed to El Arrouch, Algeria. Soon after, two more General Hospitals were despatched to Italy. On 6 November, en route to Naples, the troopship SS *Santa Elena*, which was carrying No. 14 Canadian General Hospital, was attacked by German aircraft. Struck by a torpedo and then a bomb, those aboard her took to lifeboats. Fortunately, there was no loss of life. The transport *Monterey* retrieved the 1,870 Canadian troops and nurses traveling in *Santa Elena*. Her crew and Armed Guard returned to the ship which was still afloat. However, *Santa Elena* suffered further damage the next day when accidentally rammed by the Dutch transport *Marnix Van St. Aldegonde*. Both later sank from progressive flooding, and four of *Santa Elena*'s merchant crew perished.[11]

The only Canadian nurse to die owing to enemy action during the war was a Navy sister, Sub-Lt. Agnes Wilkie. In 1942, Wilkie was the assistant matron of the Navy Hospital at HMCS Avalon in St. John's, Newfoundland. She was returning from two weeks of leave in Canada when she boarded the ferry S.S. *Caribou* in October to travel back to Newfoundland. Wilke was accompanied by her colleague and friend Nursing Sister Margaret Brooke on board the passenger and train ferry.[12]

SS *Caribou* operated in the Cabot Strait between Port aux Basques, Newfoundland and North Sydney, Nova Scotia. On 14 October 1942, the German submarine *U-69* (Kptlt. Ulrich Graf) torpedoed and sank the ferry, causing the worst loss of life in Canadian waters during the Second World War. The vessel's master, 30 crewmembers, 57 military personnel and 48 passengers were lost. The minesweeper HMCS *Grandmere* (J-258) picked up 15 of the crew, 61 military, and 25 passengers and landed them at Sydney, on the east coast of Cape Breton Island in Nova Scotia, the same day.[13]

Sister Agnes Wilkie survived the sinking of SS *Caribou* and was clinging to a capsized lifeboat with Margaret Brooke and others for hours in the frigid North Atlantic. Tragically, despite the heroic efforts of her companion Brooke, Wilkie succumbed to hypothermia before survivors were rescued by the *Grandmere*. Agnes Wilkie was buried with full naval honours in Mount Pleasant Cemetery in St. John's; and her name was given to a nurses' residence in Halifax and to Lake Wilkie in Manitoba in honour of her wartime sacrifice. Margaret Brooke was awarded membership in the Order of the British Empire, the only Nursing Sister to receive this honour during World War II. She was "Gazetted' (readers learned of her MBE in the *London Gazette*) on 1 January 1943. Her citation reads:

> For gallantry and courage. After the sinking of the Newfoundland Ferry S.S. *Caribou*, this Officer displayed great courage whilst in the water in attempting to save the life of another Nursing Sister.[14]

Thirteen days after D-Day (6 June 1944) at Normandy, which began the Allied invasion of the European continent, the first two Canadian Nursing Sisters (with No. 2 Royal Canadian Air Force Mobile Field Hospital) landed in Normandy at Bernières sur-Mer. They followed other Sisters assigned to Numbers 2, 3 and 6 Casualty Clearing

Stations set up in the Caen area nine miles from the Normandy coast. By mid-July, No. 7, 8, and 10 Canadian General Hospitals were established west of Bayeux, France.[15]

In addition to serving on land, RCAMC Nursing Sisters also served aboard hospital ships returning patients to Canada. During the Battle of the Atlantic, and combat in other theatres including the Mediterranean and Pacific, which lasted for the duration of the war, the Canadian Navy had two hospital ships, HMCHS *Letitia* and *Lady Nelson*. Both were staffed with Army sisters.[16]

CANADIAN NURSING SISTERS' THEME SONG

Affectionately referred to as "Bluebirds" for their blue and white uniforms, the Nursing Sisters of Canada have a long and impactful history with Canada's military forces. Their theme song follows:

> In my sweet little Alice Blue gown,
> When I first came to Birmingham town.
> I had had a bad trip, in a nasty old ship
> And the cold in my billet, just gave me the pip.
> We came out to nurse our own troops,
> But were greeted with measles and whoops.
> Now I'll be a granny, and sit on my fanny,
> And keep warm with turpentine stupes.
> In my sweet little Alice Blue gown,
> When I return to my home town
> They will bring out the band, give the girls a big hand,
> Being a nurse in the force, I'll be quite renowned.
> And I'll never forget all the fun,
> That I had, since I joined Number One
> I was happy and gay, to have served with MacRae
> In my sweet little Alice Blue gown.[17]

Note that "MacRae," in spite of the different spelling, in the penultimate line obviously refers to Lieutenant John McCrae of the Canadian Medical Corps author of the famous poem *In Flander's Fields* recited at all Remembrance Day services in Canada and elsewhere. McCrae died of sickness in the field on 28 January 1918.

U.S. Army Hospital Ships
Homeported at Charleston

Photo 15-1

Charleston, South Carolina, and its vicinity circa 1863.
Naval History and Heritage Command photograph #NH 2827

The U.S. Army's first three hospital ships in World War II—*Acadia*, *Seminole*, and *Shamrock*—all sailed from New York in 1943 on their maiden voyages. However, as the principal Army port of embarkation on the Atlantic seaboard, New York was already too heavily burdened with troop and cargo movements to allow its use as the home port of the Army hospital ships. Accordingly, during the fall of 1942 while the *Acadia* was still being converted at Boston into a combination trooper and ambulance ship, plans were being laid to utilise Charleston as the home port of at least one Army hospital ship.[1]

This planning proved of great value when, on 1 November 1943, these three hospital ships were assigned to the Charleston Port of Embarkation. After months of service overseas evacuating casualties from North Africa to the UK, the *Acadia* arrived at Charleston on 5

November 1943 with a full load of casualties, the first of the three to reach the new port.[2]

 Subsequently, as additional Army hospital ships were placed in service, all were assigned to Charleston. So far as is known no directive designated Charleston as the home port of the Army hospital fleet, and its function as such was long concealed in the press under the guise of "An East Coast Port."[3]

During November 1943, some 1,128 Army patients were debarked at Charleston. Subsequently, during the winter and spring of 1943-44 the port was concerned chiefly with evacuees from North Africa. The peak of the North African wounded transport was reached in May 1944 with a total of 2,581 patients.[4]

Moving ahead in the chronological sequence of this book to provide continuity in this subject, by 31 July 1944 the Army would have seventeen hospital ships in its service, with a total capacity of 9,500 beds. The remaining seven hospital ships that were to serve during the war were to be placed in service by the end of 1944; bringing the total patient capacity to some 14,600 beds.[5]

Until late in 1944 all of the Army hospital ships were based at Charleston, and all were engaged exclusively in evacuating casualties from the North African and European theatres.[6]

U.S. Army hospital ships were especially active in certain theatres, notably North Africa. USAHS *Seminole* carried 10,500 patients during the 52 months ending on 10 July 1944. Hospital ships were also used to a limited extent in support of amphibious operations, as at Salerno.[7]

A total of 460 patients were evacuated to Charleston from the UK in August, associated with the Normandy landings which launched the Allied invasion of the European continent. As a result of additional vessels in operation, the total number of patients evacuated by hospital ships to Charleston in September 1944 reached a new high of 4,273. Of this number, 1,652 were from the United Kingdom and 2,621 were from North Africa.[8]

These figures do not adequately represent the activity of U.S. Army hospital ships in the European Theatre, since many ships remained overseas for months at a time engaged in local operations between Mediterranean ports such as Naples and Oran.[9]

16

British Hospital Carrier *St. David*

Hospital carrier ST. DAVID sunk by enemy action. LEINSTER and ST. ANDREW damaged by near misses. Temporary repairs in hand. Expected ready for emergency service by 27th. S.T.L.O. Naples reporting casualties.... Please request Swiss Government to make to German Government following communication at earliest possible moment, (begins): H.M.G. have learned that three British hospital ships have been attacked by German aircraft today off Italian coast. (details).

H.M.G. protest most strongly against these attacks on hospital ships which were fully marked in accordance with the International Conventions. They request that the German authorities will issue most immediately categoric instructions to all naval and air forces operating under their command to secure full immunity from attack of British and Allied hospital ships. (Ends.)

—Admiralty War Diaries 25 January 1944 entry concerning German
air attacks on three British hospital ships at Anzio, Italy, resulting
in the sinking of *St. David*, and damage to *Leinster* and *St. Andrew*.
H.M.G (His Majesty's Government) protested these attacks to
the German Government via the Swiss Government.

Photo 16-1

U.S. Fifth Army troops wade ashore from the British infantry landing craft
HMS *LCI-281* during the first day of landings, near Anzio, on 22 January 1944.
National Archives photograph #SC 185796

On 22 January 1944, British and American troops landed on the Anzio beachhead against weaker than expected opposition. However, the German high command soon surged additional troops into the area, and their stubborn defence pinned the Allied forces on the beachhead for four months stalling hopes of a rapid advance toward Rome.[1]

The U.S. Army's 33rd Field Hospital and the 95th and 96th Evacuation Hospitals landed with assault forces and quickly set up operations. Approximately 200 American nurses were assigned to these units. On 24 January, the first bombs fell near the medical facilities. That night, the British hospital carriers HMHS *St. David*, HMHS *St. Andrew*, and HMHS *Leinster* were attacked by Luftwaffe aircraft while evacuating casualties from the beach. The *St. David*, with 226 medical staff and patients on board, received a direct bomb hit and sank. Two U.S. Army nurses on board were among 130 survivors rescued by *Leinster*. One of these American nurses, Lt. Ruth Hindman, had earlier survived the bombing and sinking of HMHS *Newfoundland*.[2]

PRE-WAR SERVICE OF THE IRISH FERRY *ST. DAVID*

The 2,700-ton Irish sea ferry *St. David* was built in 1932 by Cammell Laird at Birkenhead, an English shipyard on the west bank of the River Mersey, opposite Liverpool. She was one of a pair of new passenger vessels for the Great Western Railway; the other being *St. Andrew*. *St. David* was set to start providing service between Fishguard, England, and Rosslare, Ireland, across St. George's Channel.[3]

Map 16-1

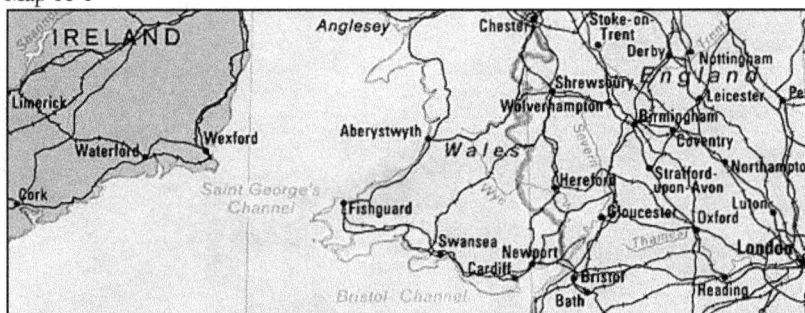

Rosslare lies about 5 miles down the coast from Wexford, Ireland

LUFTWAFFE ATTACKS AT DUSK ON 24 JANUARY

> *Enemy air opposition was heavy and continuous. The coverage given by our own Air Force was excellent but it is the writer's opinion that no amount of air coverage will completely prevent some enemy planes reaching their objective if the attacks are persistent, large in number of aircraft, and repeated frequently. The dusk attack on 24 January was a particularly vicious one in which the weapons used were dive bombers, torpedo planes, and glide bomb aircraft. It lasted something over two (2) hours.... The U.S.S. PLUNKETT and U.S.S. MAYO were seriously damaged during the attack; the hospital ship ST. DAVID was sunk, the hospital ship LEINSTER damaged; and the hospital ship ST. ANDREW attacked with unknown results.*

> —Commander Task Group 81.8 and Commanding Officer USS *Brooklyn*, Action Report – Establishment of Beachhead at Anzio, Italy by combined U.S.-British Amphibious Force – Period 21 January – 8 February 1944, 14 February 1944.

On the afternoon of 24 January, the minesweeper USS *Sway* (AM-120) was engaged escorting infantry landing craft towing a pontoon in the "swept channel" (safe water cleared of any mines present). At 1743, after a Red Alert warned ships in the area of an impending enemy air attack, *Sway* set General Quarters, ordering her crew to their battle stations. Coming under attack by German Ju 88 dive bombers, she opened fire with all guns that could be brought to bear on targets, while manoeuvering to avoid dropped bombs.[4]

At 1755, a Ju 88 crashed close aboard her bearing 075°T. Five minutes later, an unidentified plane crashed at 268°T. At 1805, *Sway* shot down a Ju 88 with her port 40mm guns. Soon after, enemy aircraft began dropping flares to provide illumination in the fading daylight; then still more bombs were dropped, as expected:

- 1830 Large group of flares on starboard quarter
- 1914 Flares dropped on starboard bow
- 1918 Flares dropped on port bow
- 1920 Flares dropped on all bearings[5]

Commanding officer, USS *Sway* (Lt. Comdr. Herman S. Strauss, USNR) reported in his action report that at 1920, following the entry "Flares dropped on all bearings," occurrence of the following events:

Heavy bomb concussions felt. Hospital ship standing out channel under attack. *St. David* sunk, *St. Andrew* and *Leinster* damaged. *Leinster* picking up survivors.[6]

At 2105, *Sway* secured from General Quarters.[7]

Photo 16-2

German Junkers Ju 88A aircraft taking off, circa 1940-1943.
Naval History and Heritage Command photograph #NH 112955

Hospital carriers *St. Andrew*, *St. David*, and *Leinster* had been lying off the Anzio beaches about one mile from shore. At approximately 1730, all three ships were ordered to put to sea for the night by Rear Adm. Frank J. Lowry, USN (commander Task Force 81), embarked on board the amphibious force flagship, USS *Biscayne* (AGC-18). Night retirement by ships was common—the belief being they would be less susceptible to attack, and thereby safer, further off the assault beaches.[8]

At 1830, while in the swept channel, *Leinster* received a direct bomb hit forward, starboard side. Fire broke out, but was brought under control by crewmembers, and she remained on her course. From 1845 to 1935, continuous bombing attacks were made on her with ordnance falling very close, but no further damage was sustained apart from severe shaking. *St. Andrew* closed *Leinster* to give assistance, but was told none was required. At 1932, a deliberate low-level aircraft attack was made on *St. Andrew*, with bombs falling within 50 feet of the ship.[9]

At 1950, *Leinster* received a signal from a lifeboat that *St. David* had been sunk. After lowering water ambulances, she picked up 101 survivors, and proceeded to Naples; arriving there at 0830 on 25 January. *Leinster* had superficial damage to her superstructure, one water ambulance was destroyed, and four others damaged—but no personnel casualties. *St. Andrew* also lowered boats, retrieved survivors, and sailed for Naples. She had been severely shaken by bomb blasts, but with no casualties.[10]

AIRCRAFT ATTACK ON, SINKING OF, *ST. DAVID*

Photo 16-3

German fighter-bomber attack on the invasion force, off Anzio, on 22 January 1944. Photographed from the destroyer escort USS *Frederick C. Davis* (DE-136). National Archives photograph #80-G-223445

It is noted before the attacks heading further seaward, the three hospital ships were all illuminated, and clearly displayed their neutral markings. When attacked, the *St. David* was about 25 miles southwest of the Anzio beachhead, making 14 knots and steering a southwest course. Winds were southwesterly, with an associated heavy southwesterly swell.[11]

The enemy aircraft that sank *St. David* overflew her from starboard to port, dropping four parachute flares which illuminated the entire area. The plane then from an altitude of about 5,000 feet, dove down to almost mast height of the hospital ship and released two bombs. At 1922, she was struck by one bomb in No. 3 hold, near the after end of the promenade deck, and shuddered violently.[12]

Following this attack, the plane turned and came in on a second attack run, and two bombs exploded alongside the ship's No 2 hold. The resultant blasts breeched *St. David*'s hull, knocked out all the lights and stopped her engines. She settled rapidly by the stern, listing 20 degrees to port, which prompted Captain Evan William Owens to order, "abandon ship." Evacuation of the patients was begun. However, the water ambulances situated along her starboard side were fouled, making them impossible to lower, despite the second officer's repeated efforts.[13]

The No. 2 water ambulance on the port side, weighing nearly two tons, was lowered into the water with a large number of patients, including six stretcher cases, and some of the ship's personnel. Because of the heavy swell, it proved very difficult to release the boat from the falls, owing to the weight on the hooks and the spring clips that retained them. With time scarce, the ambulance was freed just before *St. David* sank, stern-first, a mere five minutes after the initial bomb hit.[14]

Numbers 4 and 6 ambulances, also situated on the port side, could not be released in time and both were dragged down with the sinking ship, taking an unknown number of people with them. No. 8 lifeboat, on the third-class deck, was successfully launched, and four rafts also floated clear. Chief Officer B. Howell-Mendus and 19 others were in No. 2 ambulance, circled the wreck site and picked up 17 survivors. Boat No. 8 picked up about 10 others.[15]

By 2115 all survivors were on board the other two hospital ships (101 in the *Leinster* and 58 in the *St. Andrew*). Captain Owens lost his life with 12 of his crew; 22 Royal Army Medical Corps (including 2 Nursing Ssisters from the Reserve QAIMNS, Sarah Dixon and Winnie Harrison); and 22 patients.[16]

Nurse Jane Mansley was awarded the MBE for bravery on Hospital Carrier *Leinster*. Her letter of commendation reads:

Mansley, J. Q.A.I.M.N.S.(R).
Action for which Commended.
On 20th. Feb. 1944, whilst this Hospital Carrier was proceeding from Port d'Anzio to Naples, with wounded aboard, during the period 19.20-21.00 hrs, we were dived on by enemy planes to mast height. Many patients on our upper decks became very excitable and nervous. By the calm spirit and keen devotion to duty she was largely instrumental in restoring their confidence. Her example to all was of the highest.[17]

RETURN TO ANZIO

British hospital ship duty at Anzio continued as evidenced by the following photograph of *Leinster* in the harbour there in March 1945.

Photo 16-4

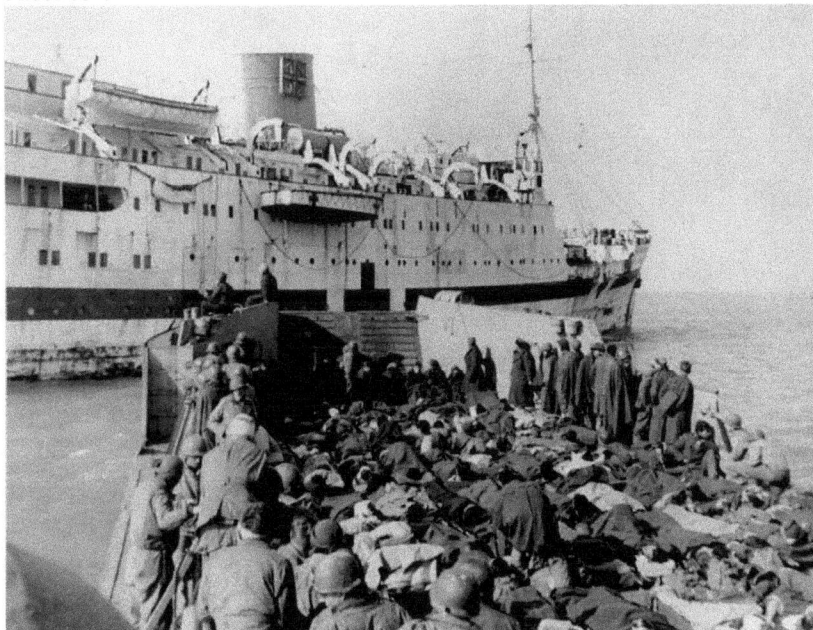

American and British casualties ride a tank landing craft as it approaches the British hospital ship *Leinster*, in Anzio Harbour, 3 March 1944. They are accompanied by hospital corpsmen and Red Cross workers.
National Archives photograph #80-G-58426

17

Normandy Invasion

The evacuation of casualties was carried out expeditiously and successfully in spite of several deviations from the planned scheme, rendered necessary by operational changes. Liaison between the medical services of the Navy and Army was good.

—"Report by The Allied Naval Commander-in-Chief Expeditionary Force on Operation Neptune, Volume I, 1944."

Photo 17-1

Normandy Invasion, 14 June 1944. Motor torpedo boat USS *PT-199* carries Adm. Harold Stark, commander U.S. Naval Forces Europe, to the Normandy beachhead. A British hospital ship lies offshore in the left distance.
National Archives photograph #80-G-253227

The Allied landings at Normandy on the coast of northwestern France, on 6 June 1944, paved the way for the liberation of Europe. Operation NEPTUNE began shortly after midnight that day with more than 2,200 bombers attacking targets along the coast and inland. Cloud cover hindered the air strikes, and the coastal bombing at Omaha Beach was particularly ineffective. More than 24,000 American, British, and Canadian airborne assault troops and 1,200 planes followed the air bombardment. At 0630, six Allied divisions and numerous small units began landing on the beaches. The Allies landed more than 160,000 troops at Normandy, of which 73,000 were Americans assaulting Omaha and Utah beaches. A combined 83,115 British and Canadian forces landed on Gold, Juno, and Sword beaches.[1]

Map 17-1

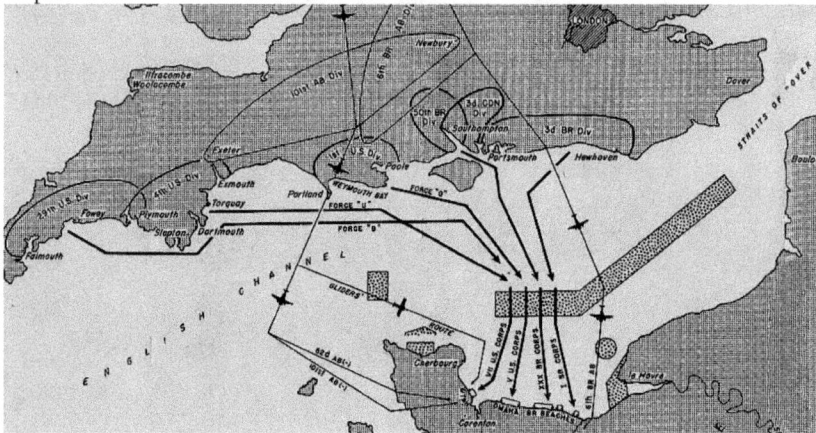

Main Allied sea routes and air routes to the Normandy beaches in the Bay of Seine. The 1st British Corps, which used the far-right approach route, included the Canadian forces that landed on Juno Beach.
https://history.army.mil/books/wwii/utah/maps/MAP2.JPG

Allied casualties on 6 June have been estimated at 10,000 killed, wounded, and missing in action: 6,603 Americans, 2,700 British, and 946 Canadians. Over the following days the Allies gradually expanded their tenuous foothold as the fighting moved inland. The Allies finally broke out of Normandy on 15 August, and once out, advanced quickly liberating Paris on 25 August. The German forces' retreat across the Seine River, five days later, marked the end of the Normandy campaign with its great loss of lives and injured bodies and minds on both sides.[2]

ROLE OF BRITISH HOSPITAL CARRIERS, SOME STAFFED BY U.S. ARMY MEDICAL PERSONNEL

On the far shore [Normandy assault beaches], First Army was initially responsible for the evacuation of casualties. After the first few uncertain days the evacuation system became well organized. A large "holding unit" was established on OMAHA Beach and another on UTAH Beach through which all of the casualties flowed. From the "holding units," casualties were evacuated either to the beach and loaded on LST's or hospital carriers, or to an airfield adjacent to the "holding unit." No such "holding unit" was available in the UTAH Beach sector for some time. These "holding units" were formed by the Medical Battalions of the Engineers Amphibious Special Brigades.

—Historical Reports and Monographs - Staff Section Reports
File Number 581-12, Surgeon General, Medical History,
ETO Vol XII, Conflict Period World War II
Roll Name MP63-9_0113.

Photo 17-2

An LCVP approaching Omaha Beach on 6 June 1944. The landing craft is smoking from a fire started by a German machine gun bullet striking a hand grenade. After discharging his troops, the boat's coxswain, assisted by the engineman and bowman, put out the fire and returned to their attack transport, USS *Samuel Chase* (APA-26). National Archives photograph #26-G-2342

Almost immediately after Allied troops began storming ashore at Normandy, beachmasters began requesting the urgent evacuation of casualties. Stretchers from LCVPs ("Higgins boats") were sent into the beach from attack transport ships (APAs) offshore, and were available at H-hour plus 60 minutes (0730). However, heavy enemy fire on the beaches greatly impeded the removal of wounded assault forces. To correct a shortage of medical supplies, resupply units were also despatched from APAs but their distribution ashore was similarly hindered by continued enemy fire.[3]

All LSTs (tank landing ships) assigned to the operation had been refitted to receive casualties and operating rooms had been provided. Hence, casualties were delivered by LCVPs ("Higgins boats") and other craft receiving them on the beach to the nearest LST for treatment.[4]

Photo 17-3

An LCVP from USS *Thomas Jefferson* (APA-30) pulls away from the shore at a southern English port with a full load of troops who will be carried to invasion shipping in the harbour, circa early June 1944.
National Archives photograph #80-G-252140

The first casualties from the far shore (Normandy) began to arrive in England on D+1 (7 June) with a heavy increase on D+3. This was due to the fact that LSTs were held on the far shore until a large convoy was formed. One of the key points for delivery of patients was the region around Weymouth where LSTs could beach at Portland (a port city and naval base on England's south coast) and other smaller craft at the docks in Weymouth (a seaside town to the west of Portsmouth).[5]

Photo 17-4

LST unloading litter patients to an LCT (tank landing craft)
off Weymouth, England, for their transfer ashore, 11 June 1944.
Historical Reports and Monographs - Staff Section Reports File Number 581-12
Surgeon General, Medical History, ETO Vol XII Conflict Period World War II

Medically-equipped LSTs evacuated a number of casualties. The
first hospital carrier did not arrive off Normandy until 7 June, owing to
mine casualties to HMHS *Dinard* and *St. Julien* (which were knocked out
of service for several weeks, until repairs were completed).[6]

HMHS *Naushon* arrived off Omaha Beach at 1900 on D+1, and
began taking patients aboard from craft lying offshore. Contrary to
plan, she remained overnight, caring for the injured by means of the
First U.S. Army Medical Detachment "A" which was aboard.[7]

Photo 17-5

Hospital carrier *Prague* off the Isle of Wight, August 1944. During the Normandy
Invasion, First U.S. Army Medical Detachment personnel provided medical care,
while her British Merchant Marine crew operated the ship.
Historical Reports and Monographs - Staff Section Reports
File Number 581-12, Surgeon General, Medical History,
ETO Vol XII, Conflict Period World War II
Roll Name MP63-9_0113

   Casualties were evacuated to hospital carriers by DUKWs and
water ambulances. Swells and risk of damage to craft from carriers'
rubbing strakes (designed to prevent damage to a ship's hull when going
pierside or when craft come alongside), as well as the distance offshore
that carriers were compelled to lie, made evacuation to hospital carriers
slower and more difficult than using ramps aboard LSTs.[8]

Photo 17-6

U.S. Army DUKW amphibious trucks bring supplies to Utah Beach from ships
anchored off shore, 8 June 1944. A German beach gun is in the right foreground.
National Archives photograph #SC 190232

Photo 17-7

Transferring wounded from a DUKW amphibious truck to a water ambulance
before hoisting to the promenade deck of hospital carrier HMHS *Prague*.
Historical Reports and Monographs - Staff Section Reports File Number 581-12
Surgeon General, Medical History, ETO Vol XII Conflict Period World War II

   Most of the hospital carriers had six water ambulances. Operated
by members of the ship's merchant crew (a quartermaster and two able

body seamen), the boats plied waters often under fire, and at one period swept by a gale, to collect patients. Each water ambulance could normally take six wounded on stretchers and a handful of other wounded sitting. This very limited capacity necessitated many trips back and forth to DUKWs/landing craft bringing wounded from the beaches out to water ambulances for further transfer to hospital carriers.[9]

The hospital carrier *Lady Connaught* arrived off Normandy the night of 8-9 June. Her rated capability was approximately 300 casualties. Aboard her was First U.S. Army Medical Detachment "B"—consisting of the following personnel:

- Station and litter bearer platoons of the 502nd and 427th Medical Collecting Companies
- 31st Medical Group
- Six surgical teams of the 4th Auxiliary Surgical Group
- One Advance Depot Platoon
- 31st Medical Depot Company
- Six Medical Corps officers from the 662nd Medical Clearing Company
- 134th Medical Group
- Ten liaison officers from various medical units, including one officer from the 9th Troop Carrier Command[10]

HMHS *Naushon*, *Lady Connaught*, and *Isle of Jersey* were among ten hospital carriers used in the opening phase of the Normandy landings; later there were twelve. Initially, some difficulty was experienced by DUKWs/landing craft carrying casualties in locating hospital carriers known to be in the area, as they were berthed too distant from the evacuation points on the beach. The carriers were moved inshore, but difficulty in locating them at night continued until arrangements were made to berth each hospital carriers on arrival in the same general area.[11]

By 10 June, it became apparent that casualties could readily be handled by one daily hospital ship, and one or two hospital converted-LSTs, retained in the area for the purpose. Accordingly, from that time forward, evacuation of casualties to the nearest LST was discontinued. Army personnel had meanwhile constructed an airfield back of the beaches and on 10 June, evacuation of casualties by transport planes commenced, lightening the load on hospital carriers and LSTs.[12]

Royal Canadian Army Medical Corps' Service, Including at Normandy

	Regular Force
	2 July 1904: Permanent Active Militia Army Medical Corps established. 1 May 1909: Redesignated Canadian Army Medical Corps. 3 November 1919: Redesignated the Royal Canadian Army Medical Corps 2 July 1974: Disbanded. Medical Service Branch of Canadian Forces formed by the amalgamation of The Royal Canadian Army Medical Corps, and equivalent services of the former Royal Canadian Navy and former Royal Canadian Air Force
Royal Canadian Army Medical Corps Badge	

During the WWII European war, 34,786 personnel served in the Royal Canadian Army Medical Corps, including 3,656 Nursing Sisters, and the Corps suffered 107 fatal battle casualties.[1]

On the front lines were personnel assigned to Field Ambulance organisations, who were responsible for the evacuation and treatment of casualties forward of Casualty Clearing Stations. Field Ambulance units were assigned to support specific brigades, for example:

- No. 14 Canadian Field Ambulance worked with 7th Canadian Infantry Brigade casualties
- No. 22 with 8th Brigade casualties
- No. 23 with 9th Brigade casualties[2]

PATIENT EVACUATION CHAIN

Assault sections of these three Field Ambulance units landed with the infantry on D-Day, launching Allied amphibious assault operations. From the battlefield, wounded soldiers were moved by stretcher-bearers to their unit's Regimental Aid Post (RAP), from which they were evacuated by ambulance.[3]

RAPs were set up in haste to deal with the wounded as quickly as possible, so only very basic treatment was available. They were sometimes bypassed, and casualties taken directly to Casualty Clearing Posts (CCPs), where they might receive blood products or morphine. The entire chain of evacuation to CCPs was within range of enemy fire, so removal of casualties further to the rear as quickly as possible was obviously of extreme importance.[4]

The next step was evacuation to a Field Dressing Station, where intermediate treatment was available before transfer to a Casualty Clearing Station, a basic hospital offering surgery and short-term convalescence.[5]

After Normandy, this system was modified in Northwest Europe. In order to get the wounded into surgery faster, Army medicine combined Field Dressing Stations, Field Transfusion Units (FTUs) and Field Surgical Units (FSUs) to form Advanced Surgical Centres (ASCs). Casualty Clearing Stations (CCSs), which otherwise performed surgery, were not considered suitable to function as ASCs because they had insufficient personnel to adequately support two FSUs (each with two surgical teams). The ASC operated closer to the front while the CCS came to be primarily responsible for convalescence further back.[6]

UTILISATION OF MEDICAL PERSONNEL

Treatment of the wounded in forward areas was the responsibility of male medical personnel, but the contribution of nursing sisters to post-operative care of wounded soldiers cannot be understated. Nursing sisters were usually attached to a General Hospital or CCS, but arguments were made for their employment further forward with FSUs because their role in monitoring patients following surgery not only aided recovery, it also made the surgeon's job easier.

—From the article "Canada in the Second World War, The Army Medical Organisation" by the Juno Beach Centre, 2004.[7]

The final step in the evacuation chain was patient transfer to a General Hospital for cases requiring further care. In the early stages of the Normandy invasion, this was done primarily by tank landing ships and hospital ships to England, although air evacuation began once an air strip was built behind the assault beaches.[8]

Photo 18-1

Tank landing ship *HM LST-427*'s tank deck with some of the 456 casualties on board being Medivaced to England, date unknown.
Courtesy of NavSource and Robert Hurst

By July 1944, a number of Canadian General Hospitals were concentrated at Bayeux, Normandy, as part of the 21st Army Group Medical Centre, and No. 6 Canadian General Hospital had moved to Douvres-la-Délivrande in northwestern France. As the Allied armies advanced through France and Belgium, Canadian hospitals moved to the Rouen and Dieppe areas, later to Antwerp and Germany.[9]

CAUSES OF CASUALTIES / MORTALITY RATES

The primary killer of Canadian forces throughout the war was artillery, as shown by the following summary information:

- Artillery: Nearly 43 percent
- Machine guns: 17.5 percent
- Mortars: 14.5 percent
- Rifles: 10 percent[10]

The percentages add up to 85 percent, which suggests that they were identified by bullet, or artillery and mortar shrapnel wounds, and the causes of the remaining 15 percent could not be determined. As noted elsewhere in the book, casualties of explosive blasts typically had no distinguishable wounds, and instead internal fatal organ injuries.[11]

Of significance were advances made in battlefield medicine. The mortality rate of World War II wounded dropped to 66 per thousand from 114 per thousand in the Great War. Moreover, as a result of improved sanitation, hygiene, and treatment methods, less than 1 percent of fatalities were due to disease.[12]

19

British Hospital Carrier *Amsterdam*

Photo 19-1

Hospital carrier HMHS *Amsterdam* under way, location and date unknown.
https://www.harwichanddovercourt.co.uk/wp-content/uploads/2012/04/HMHS-Amsterdam.jpg

The British hospital carrier HMHS *Amsterdam* (the 8th and last British hospital ship/carrier of the war to be lost in the European Theatre) was sunk off Juno Beach, Normandy, on 7 August 1944 by a German mine.

The 350-foot *Amsterdam* was built in 1930 by John Brown's Shipyard, Clydebank, Scotland, as a passenger ferry for the London & North Eastern Railway Company. On 12 September 1939, she was requisitioned as a troop ship by the Ministry of War Transport. She was used in May 1940 for the evacuation of British troops from France and the repatriation of evacuated French troops. (Of the French soldiers evacuated from France to England, some chose to join Charles de Gaulle's Free French army in Britain, while others returned to France to fight the Germans.) In February 1941, *Amsterdam* took up Orkney and Shetland's troopship service.[1]

In January 1944, *Amsterdam* arrived in North Shields, a port town eight miles northeast of Newcastle in North East England. There, she was fitted out as an infantry landing ship capable of carrying six assault landing craft (LCAs) and 420 troops. Patrick Manning, an underage galley boy (he forged his papers) who joined *Amsterdam* in February, described the atmosphere aboard the infantry landing ship, and her early operations for the Normandy Landings:

Nearly all the crew came from Harwich and Lowestoft, having been with the ship since the days before the war, when it was a regular passenger ship sailing from Harwich to the Hook of Holland. They were nearly all related to each other, fathers, sons and brothers and it created a family atmosphere on board. The ship had been camouflaged and carried six LCAs, three on each side on davits. Every week from February till June we would carry American Rangers round to Swanage [a port town in southeast England], where we would anchor off the cliffs. The troops would then go ashore in the boats and scale the cliffs for practice. In between times we would go out in the Channel and pick up people in small boats that had come across from France.[2]

On 5 June, *Amsterdam* took part in Operation NEPTUNE (the Normandy invasion), leaving from Weymouth Bay as part of Assault Convoy O, carrying U.S. Army Rangers to Omaha Beach.[3]

Photo 19-2

U.S. Army Rangers show off the ladders they used to storm the cliffs at Pointe du Hoc, during the Normandy Invasion, which they assaulted in support of Omaha Beach landings on D-Day, 6 June 1944. Photograph released on 12 June 1944.
National Archives photograph #80-G-45716

Following this duty, she proceeded to Glasgow where she was converted into a hospital carrier and fitted with six water taxis. Manning recalled observing the Rangers scaling Normandy cliffs, and *Amsterdam*'s subsequent conversion to a hospital ship:

> Eventually 'D' Day dawned we arrived off the coast of France around 3.00 a.m. Our LCAs, loaded with U.S. Rangers, left the ship and headed towards the cliffs and Omaha beach. When they arrived, we could see them firing the rockets up with ropes attached to them with hooks so they caught on the rocks. We then watched as the [battleship] USS *Texas*, [light cruiser] HMS *Belfast* and what I think was [light cruiser] HMS *Glasgow* bombarded the coast. The Rangers lost a lot of men, and two of our LCAs, were sunk.
>
> We left the French coast that night and sailed to Glasgow…. [There the ship was] completely refitted out as a hospital ship containing wards, operating theatres etc. The crew's quarters had been changed and all the catering staff were put in cabins amidships. There was not enough room for my friend and myself, so we were put in a small cabin right at the aft end of the ship. We didn't think much of this at the time but, as it turned out, it was a blessing in disguise.[4]

Photo 19-3

Glasgow, Scotland, 1940. Damage to shipping resulting from a German bomber raid. Australian War Memorial photograph P00090.098

On 19 June, *Amsterdam* left for the Normandy coast to pick up her first casualties. Manning described the subsequent Southampton to Normandy and back cross-Channel trips bringing wounded personnel back to England:

> We sailed back to France and for the next few trips we ferried the wounded back to Southampton. We would anchor off the French coast, opposite the town of Caen, our LCAs, would go ashore, and pick up the wounded from the field hospital, then we would return to Southampton.[5]

Photo 19-4

Miracle Harbour 8 (Normandy) pierhead in action. Wounded are being transferred from the ambulances to a hospital ship for return to England.
Naval History and Heritage Command photograph

On 7 August on her third return trip, one hour off Juno Beach bound for England with 258 patients, she hit a mine and sank in 11 minutes. Of the 419 people on board, fifty-five wounded men were lost as were ten medical staff, thirty crewmembers, and eleven German prisoners of war being transported to a POW camp in England.[6]

Manning, who was asleep at the time, was able along with his cabin mate, to escape the ship before she slipped beneath the water's surface:

> [Steward Artie Mallows] gave us a call in the morning, but we fell asleep again. Suddenly there was a muffled explosion, the lights went out, and the ship listed. We managed to get our trousers on and our lifejackets, but as we looked out of the porthole, all we could see was water and the deck was wet underfoot. There was a horrible smell of ether in the air. We found the cabin door but couldn't open it. In desperation we kicked and banged on the door but nothing happened. We then noticed that a piece of one of the bunks was jammed against it. This was all happening within minutes but it seemed like forever. It was a nightmare.
>
> We managed to get out into the alleyway and made our way to the stairway. It was there we found Artie Mallows…standing by the stairs with blood running down his face. He had been in the stores when the ship was hit and must have been hit by boxes falling on him. We found him a lifejacket and helped him on deck.
>
> It was approximately 7.00 a.m. and foggy; the ship seemed to be broken in the middle with one half listing one way and the other half the other. One of the funnels and the mast were down and the screws were out of the water. Only one LCA could get away to pick up survivors.
>
> There seemed to be only us three on the aft end of the ship, when we heard someone shouting. We told Artie to stay where he was and scrambled up to the after end. There was a sergeant in the medical corps standing there, ready to jump over the side. If he had he would almost certainly have hit the propellers and died. We managed to take him back to where we left Artie, but Artie had vanished. We found the sergeant a lifejacket and got him over the side to a ledge and then jumped into the water. There was quite a heavy swell so, finding it hard to swim in a lifejacket, I paddled.
>
> I kept thinking to myself that I had to get away quickly in case I was sucked down with the ship. Then I spotted the LCA. As I reached it I was told there was no room as they were already overloaded and

were afraid of capsizing. I reluctantly let go of the grablines and carried on paddling.

I could hear a lot of screaming and shouting. I looked around and could see some of the wounded soldiers jumping over the side, and there were two people stuck in portholes. I was told afterwards that they were nurses.

As I continued paddling, I heard a gushing noise then saw what looked like ashes shooting out of the funnel amid lots of noise. Then there was nothing, just wreckage floating in the water and deathly silence.

I continued paddling. It seemed to get foggier and at one point I thought I was sinking.   After what seemed like hours I heard engines; I shouted and shouted and an American torpedo boat loomed up.   I was taken to a naval cruiser and from there to a hospital ship that brought us home.[7]

HEROIC ACTIONS OF NURSING / MEDICAL STAFF

Following the mine strike to *Amsterdam*, scant time was available for nursing staff to get as many wounded personnel from below decks into lifeboats before the hospital ship sank. With the ship's listing, resulting in the deck being angled to the surface of the water, Sisters Dorothy Anyta Field and Molly Evershed bravely returned to the lower decks. Together they rescued 75 men, without any thought to their own safety, before, sadly, the *Amsterdam* sank taking them with her.[8]

Sisters Field and Evershed were posthumously awarded the King's Commendation for Brave Conduct which could be awarded to both military and civilians during war and peacetime, and in non-military circumstances (*London Gazette* of 29 December 1944).[9]

Following are details of other QAIMNS and RAMC awards in recognition of gallant and distinguished services for their actions during the ship sinking with patients aboard.

The most excellent Order of the British Empire.
To be Additional Members of the Military Division of the said
Most Excellent Order: Sister Miss Elsie Roberts (206427)
Queen Alexandra's Imperial Military Nursing Service

This nursing officer showed a most valuable example in assisting the evacuation of patients from the wards to the promenade deck of Hospital Carrier *Amsterdam* when the ship was sunk by enemy action off the coast of Normandy on 7th August, 1944. In emergency stations drill she had been shown that her place was officer in charge No. 2 water ambulance; she continued, however, to help patients up the deck, after the water ambulance had been lowered and when the ship was actually capsizing. Miss Roberts escaped down the almost horizontal side of the ship just before it sank. Her calmness was an important factor in averting panic and by her aid many patients reached the deck who would otherwise have been trapped in the wards.

Sister Miss Ellen Theresa Hourigan (305029)
Queen Alexandra's Imperial Military Nursing Service Reserve

This nursing officer displayed much courage and a high sense of responsibility for the safety of her patients when Hospital Carrier *Amsterdam* was sunk by enemy action off the coast of Normandy on 7th August, 1944. She was aware that, in Emergency Stations, she was officer in charge No. 4 water ambulance. Nevertheless, Miss Hourigan continued to help her patients to escape to the promenade deck from B deck, of which she was jointly in charge. By her calm example, as well as by helping the more severely wounded patients who were trying to climb the stairs, she reduced the possibility of panic. Miss Hourigan did not leave the ship until after it capsized, when she escaped by scrambling down the almost horizontal starboard side, just before it sank.

Sister Miss Lily McNicholas (246129)
Queen Alexandra's Imperial Military Nursing Service Reserve

On the 7th August, 1944, during the sinking by enemy action of Hospital Carrier *Amsterdam* off the coast of Normandy, Miss McNicholas rendered important service whereby lives were saved which otherwise would have been lost. Well knowing that her place in emergency stations was in No. 3 Water Ambulance as officer in charge, she continued to give encouragement and help in bringing patients from her wards on A deck up to the promenade deck. Her patients were admitted to the ship as stretcher cases, but in the grave emergency all, save the quite helpless were obliged to walk. In this distressing situation, Miss McNicholas' presence and her cheerful encouragement was of great value. She left the ship after it had capsized by scrambling down the starboard side which was almost horizontal by that time. Unable to swim, she had much difficulty and felt ill in the water, where she was helped by the Master of the ship and by an officer of the Royal Army Medical Corps. Shortly afterwards she was rescued by an American cutter. With complete disregard of her own comfort she immediately rendered aid to patients as they were rescued from the sea.

Lieutenant Oliver Gray, (305170), Royal Army Medical Corps

Lieutenant Gray displayed great courage and devotion to duty when Hospital Carrier *Amsterdam* was sunk by enemy action off the coast of Normandy on 7th August, 1944. He was in charge of two wards on the lowest deck of the ship and evacuation required great steadiness. Moreover, one ward, through which the ship broke in two 14 minutes after the explosion was under his charge. He superintended the evacuation of patients and did not leave these wards until all living patients had gone. The ship capsized and sank while this officer was standing on her horizontal starboard side in the endeavour to rescue a nursing officer through a port hole of the lower deck in which she had been trapped.[10]

EMPLOYMENT OF HOSPITAL SHIPS ON SAME DAY

On the same day that *Amsterdam* was lost, 7 August, other hospital ships were busy evacuating casualties from France. To the west of the Normandy beaches, the Battle of Cherbourg had taken place 6 June-1 July 1944, involving the capture of the Cotentin Peninsula and, most importantly, the port city of Cherbourg with its valuable harbour.

Photo 19-5

Wounded Americans are hoisted aboard a British Hospital Ship from a tank landing craft, en route to Britain from the combat zone in Northern France, 7 August 1944. Location is probably Cherbourg. Casualty in lower left centre is reading a copy of the *Stars and Stripes* newspaper.
National Archives photograph #SC 192645

Invasion of Southern France
– Operation DRAGOON

In August 1944, we joined in an amphibious assault force to land in the South of France, on the French Riviera. We left from North Africa, and strangely enough we took French troops that had left France in 1940 when the Germans pushed them out of France, so they were going back to their own homeland. We also took some Americans and we operated from North Africa to the South of France and it was to a place called St. Raphael on the French Riviera. It was called the Champagne landing this one, I think, there was very little opposition because the Germans had pulled back and when we were on the beach some of the lads, some of the sailors went up and picked grapes out of the vineyards, and it was one of the easier or better operations that we had.

—Alwyn Thomas, crew member of the tank landing ship HMS *Bruiser* (F127).[1]

Map 20-1

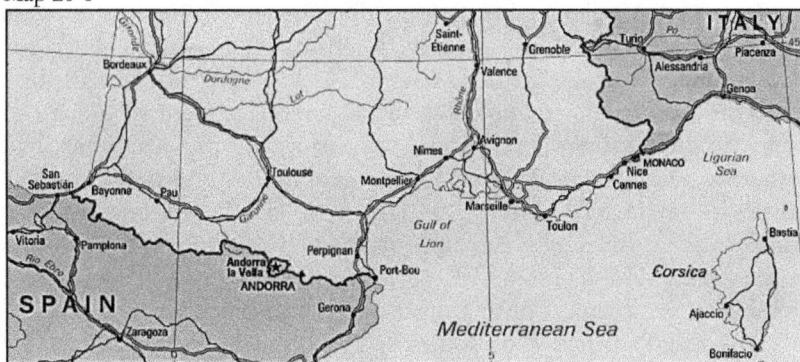

Southern France and adjacent areas

In August 1944, two months after the Normandy landings in northwestern France, an Allied amphibious assault was made on the southern coast of France between Hyeres (located about ten miles east of Toulon) and Cannes. The objectives of the operation were to secure beachheads, capture neighbouring ports, and strike north up the Rhone valley and finally to link up with Patton's Third Army. This would form the right wing of the Allied force invading Germany, and cut off German forces in the west of France. During its early planning, the code name initially given was Anvil to complement Sledgehammer, code name for the invasion of Normandy.[2]

Subsequently, Sledgehammer became Operation OVERLORD and Anvil, DRAGOON. Reportedly Winston Churchill, who opposed the plan and claimed to having been "dragooned" into accepting it by Roosevelt and Eisenhower, chose the latter name. The British prime minister believed that the operation would divert military resources that could be employed supporting ongoing operations in Italy. He preferred, instead, an invasion of the oil-producing regions of the Balkans. Churchill reasoned that by attacking the Balkans, the Allies could deny Germany oil, as well as hinder the advance of the Red Army of the Soviet Union—thus achieving a superior negotiating position in post-war Europe with that nation.[3]

When the invasion of southern France was initially considered, the Allied landing at Anzio had gone badly, and hence planning for the French one was "put on hold." The operation—renamed DRAGOON as previously mentioned—was revived, following the successful Normandy invasion of 6 June 1944 (D-Day). In Italy after the capture of Rome on 5 June, it was possible to allocate troops to the invasion of southern France but it was not possible to obtain the requisite assault shipping, concurrently with the Normandy invasion.[4]

It became available shortly after the landings at Normandy were completed. There was urgency to bring these ships into the Mediterranean to join the U.S. Eighth Fleet so that additional ports could be captured to increase supply, as the Allies were struggling to resupply their armies in France. Considering the Germans had destroyed the port facilities at Cherbourg, and a violent storm had damaged the artificial harbour "Mulberry A" off the American forces assault beach Omaha at Normandy, it was important to seize the ports at Marseilles and Toulon. Also, Allied Free French leaders were pressing for an invasion of their homeland from the south.[5]

The assault beaches of southern France were well protected by German coastal defences. However, enemy fortifications lacked depth inland from the beaches and there were few reserves for counterattack. Allied planners chose the Cavalaire Bay-Rade d'Agay area for the assault because it was within range of supporting Allied aircraft operating from fields in Corsica. It offered a favourable sea approach with only a narrow coastal area suitable for enemy mining, and it had the fewest German coastal gun batteries capable of reaching the approach, transport areas, and landing beaches. Further, establishment of a bridgehead in this area would provide a suitable base for a landward attack on Toulon and Marseilles, and rapid movement up the Rhone Valley in southeastern France.[6]

ALLIED ASSAULT FORCE

The naval force for Operation DRAGOON was under the command of Vice Adm. Henry K. Hewitt, USN, commander Eighth Fleet and commander Western Naval Task Force. It included 503 U.S. ships and craft, 252 British, 19 French, 6 Greek, and 63 merchant ships of various nations—a total of 843 vessels. Shipborne on this fleet were 1,267 landing craft.[7]

Photo 20-1

Shipping in the bay of Naples, Italy, circa early August 1944, prior to departing for the southern France invasion. Two Italian light cruisers are present, along with dozens of U.S. Navy and British amphibious and combatant ships.
National Archives photograph #80-G-59468

The Allied Assault Force comprised three divisions of the U.S. Seventh Army—assault force Alpha (U.S. 3rd Infantry), Camel (U.S. 36th Infantry), and Delta (U.S. 45th Infantry) under their respective commanders—and two divisions of French Army B under the command of Gen. Jean de Lattre la Tassigny.[8]

LANDINGS OF TROOPS A GREAT SUCCESS

On D-Day (15 August) the three U.S. infantry divisions were successfully landed in force in southern France. The 3rd Infantry came ashore on beaches in Cavalaire Bay and Pampelonne Bay and moved rapidly inland, with leading elements reaching Grimaud, 6 miles west of St. Tropez. The 45th Division landed on beaches in Bougnon Bay, with leading elements reaching Plan de Tour, 4 miles northwest of St. Maxime that day. Landing over beaches east of Frejus, the 36th Division fanned out rapidly to the north, east, and west, reaching the high ground overlooking Frejus from the northeast.[9]

In a precursor action, the 1st Special Service Force was landed on Ile de Levant and Ile de Port Cros to neutralize coastal gun batteries on the islands. The French Groupe de Commandos was landed near Cap Negre and occupied the high ground, repulsing a German counterattack during the day. After landing south of Theoule-sur-Mer, the French Naval Assault Group made contact with the U.S. 36th Division.[10]

HOSPITAL SHIP CARE OF WOUNDED SOLDIERS

Plans for seaward evacuation of casualties were influenced by experiences gained in previous opeations in this theatre. The Mediterranean is peculiar in that the distances between friendly shores and target areas are short enough to provide quick turn around times for hospital ships but at the same time they are too long to make routine use of LSTs desirable for the evacuation of litter cases.

The army was responsible for the care of casualties landward of the high water mark, for their transportation to the navy evacuation stations and for the furnishing of a check-off list of all casualties so transported. The navy was responsible for the care of casualties in the navy evacuation stations, for evacuating casualties seaward and for keeping a record of all casualties so evacuation.

—Vice Adm. Henry K. Hewitt, USN, describing the clear-cut division of responsibility between U.S. Army and Navy medical sections incorporated in his medical plan for the invasion of southern France.[11]

During high-level planning for the invasion of southern France, it was recognised that location of U.S. Army and British hospitals in Naples, Italy, and French Army hospitals in Oran, Algeria, would make it necessary to evacuate their respective casualties to these ports. It was estimated that a minimum of 12 hospital ships would be needed to evacuate the maximum number of casualties from the combat area to hospital in the rear echelon, but 15 would be an ideal number.[12]

Vice Adm. Henry K. Hewitt, USN (commander, U.S. Eighth Fleet, and Naval Commander, Western Task Force) requested 15 hospital ships. The U.S. War Department allocated 12 to the operation because of pressing requirements elsewhere; the European and Pacific theatres had top priority for American hospital ships and the British Eighth Army and Middle East Theatre for the British ones. The Navy hospital ship USS *Refuge* (AH-11) was later made available as a replacement for one of twelve allocated by the Army, which was delayed in her overhaul and conversion in the United States.[13]

The twelve hospital ships were placed in a Hospital Ship Pool under the operational control of the Royal Navy Principal Sea Transport Officer of Allied Force Headquarters. They were sent into the combat area on the planned schedule shown in the table, beginning on D plus 1 and continuing through D plus 6.

Staging of Hospital Ships to the Combat Area off Southern France (Eleven U.S. Army and One U.S. Navy)

Date	Ship	Patient Capacity	Speed
D+1 (16 Aug 44)	USAHS *Algonquin*	455	14.5
	USAHS *Chateau Thierry*	511	15.3
	USAHS *Shamrock*	543	14.5
D+2 (17 Aug 44)	USAHS *John L. Clem*	291	14.5
	USAHS *Acadia*	788	18
	USAHS *Thistle*	462	15
D+3 (18 Aug 44)	USAHS *Emily H. M. Weder*	737	13
D+4 (19 Aug 44)	USAHS *Marigold*	799	13
	USAHS *Ernest Hines*	287	14
D+5 (20 Aug 44)	USAHS *Seminole*	456	13
	USAHS *John J. Meany*	579	11
D+6 (21 Aug 44)	USS *Refuge*	630	10.5[14]

All landing craft transporting troops and equipment from ship to shore were available to evacuate casualties from shore to ship. Navy Beachmasters (responsible for naval operations in the vicinity of the beaches) were directed to use these craft whenever possible. Fortunately, casualties were only a fraction of those anticipated, and small boat personnel performed excellent work in transporting wounded to evacuation ships.[15]

Hospital ships were not ordered into the combat zone on D-day, because experience had shown assault day casualties to be relatively lighter than those of succeeding days, and facilities afforded by troop transport ships were believed sufficient to meet those requirements. Twelve transport ships took park in the operation and were equally distributed among the three task forces. Their Navy medical staffs were augmented by personnel in the theatre, thus permitting them to use their medical department facilities to capacity. D-day casualties were very light on all beaches and the transports were more than sufficient to meet the requirements.[16]

As planned, beginning on D plus 1, hospital ships arrived at the outer screen of ships off the assault beaches at sunrise on assigned days, and were directed to proceed to anchorages off the beaches where casualties had been collected. Embarkation of patients continued until one hour before sunset when the ships departed the area. Hospital ships loaded to less than 50 percent were directed to proceed to the other screen, and to return the following day.[17]

D plus 7 (22 August) marked a new method of hospital ship utilisation, one that succeeded the previous planned schedule. On this date, the RN Navy Principal Sea Transport Officer began despatching hospital ships to the combat area only upon request of the U.S. Seventh Army. Army units ashore had by then established sufficient Evacuation and Field Hospitals ashore to give them a satisfactory holding capacity.[18]

The schedule used up to that point had functioned very satisfactorily. During its operation, approximately, 1,800 casualties had been evacuated from the assault beaches to hospital ships.[19]

Attempts were not made during the early phases of the invasion of southern France to classify casualties by nationality on the beaches. Instead, most patients were evacuated to Naples where classification occurred. From there French casualties were then sent to Oran as transportations became available. Later the French Army was able to use several civilian hospitals which were found functioning in Toulon and Marseilles. French casualties taken to these facilities did not want to leave. Consequently, the numbers of wounded soldiers on beaches awaiting evacuation to Naples were materially reduced.[20]

Later, when it became necessary to maintain availability of a sufficient number of beds in the invasion area for new casualties, these local beds were cleared with their French occupiers taken to hospital ships for transport to Oran. It is noted a small number of French casualties were first evacuated to Naples, before transport to Oran. All American and British casualties were evacuated to Naples.[21]

The remainder of the chapter is devoted to the USS *Refuge*. During World War II, the Navy's hospital ships were employed almost exclusively in the vast Pacific Theatre. However, as previously mentioned, *Refuge* supported the invasion of southern France only because replacement was needed for an Army hospital ship unable to complete overhaul/refurbishment on time.

NAVY HOSPITAL SHIP *RUFUGE*

Photo 20-2

Hospital ship USS *Refuge* (AH-11) at Baltimore, Maryland, March 1944.
National Archives photograph #80-G-170967

The 522-foot hospital ship USS *Refuge* (AH-11) had been built in 1921 by the New York Ship Building Corporation of Camden, New Jersey, as the passenger liner SS *President Madison*. Owned and operated by the American President Lines, she was acquired by the Navy from the War Shipping Administration on 11 April 1942 for conversion to a troop transport. Then named USS *Kenmore* (AP-62), she was placed in commission at Baltimore, Maryland, on 5 August 1942, under the command of Comdr. Myron T. Richardson, USN, and subsequently served in the Pacific.[22]

On 16 September 1943, she was placed out of commission at Baltimore, Maryland, for conversion to a hospital ship by the Maryland Dry Dock Company. After conversion and renamed USS *Refuge* (AH-11), she was placed in commission at Baltimore, on 24 February 1944 under the command of Comdr. M. A. Jurkops, USNR. *Refuge* was partially fitted out at Baltimore, and received remaining equipment at the Norfolk Navy Yard. Assigned to the Service Force, U.S. Atlantic Fleet, her mission was to transport casualties from the various war zones to the United States.[23]

Refuge departed Hampton Roads, Virginia, on 20 April 1944, bound for Mers-el-Kebir, Algeria. After embarking patients there on 6-8 May, she returned to Charleston, South Carolina. Subsequently, she made two voyages to the British Isles during the period 1 June to 29 July 1944, embarking patients at Belfast, Ireland; Liverpool, England; and Milford Haven, Wales. These patients were returned to Newport News and Norfolk, Virginia.[24]

On 2 August, *Refuge* again sailed for the Mediterranean, arriving at Oran, Algeria, on 17 August 1944. The following day, she proceeded to the coast of southern France and from 22 to 30 August made two voyages between St. Tropez Bay and Naples, Italy, evacuating casualties from the southern France invasion. On 16 September she embarked casualties from Naples and departed for Oran where she took aboard more patients, then steamed for New York, arriving 6 October 1944.[25]

Photo 20-3

Ens. Ernestine Hess, assigned nursing duties on board USS *Refuge*, posing in front of the hospital ship's stack. National Archives photograph #80-G-045037

Converted Liberty Ships

EC2 ("Liberty Ship")-The slow, all cargo type, "plow-horse" of the war. Over 2,600 constructed, of which many were converted to special uses. Reciprocating engine. Overall length, 442 feet; speed, 11 ½ knots; horsepower, 2,500.

—Roland W. Harris, *Troopships of World War II*, 1947

Photo 21-1

Hospital ship USAHS *Dogwood* in World War II.
Naval History and Heritage Command photograph #NH 103121

Liberty Ships Converted to U.S. Army Hospital Ships

U.S. Army Hospital Ship	Former Liberty Ship	Built	Initial USAHS Service
USAHS *Blanche F. Sigman*	SS *Stanford White*	1943	Jul 1944
USAHS *Dogwood*	SS *George Washington Carver*	1943	Jul 1944
USAHS *Jarrett M. Huddleston*	SS *Samuel F. B. Morse*	1942	Sep 1944
USAHS *John J. Meany*	SS *Zebulon B. Vance*	1942	Jul 1944
USAHS *St. Olaf*	*St. Olaf* was original name	1942	Aug 1944
USAHS *Wisteria*	SS *William Osler*	1942	Jul 1944

In 1943 before and after the U.S. Army's first WWII hospital ship USAHS *Acadia* was put into service, the Army had little choice in the vessels available for conversion into these vessels. Any troop lift ships lost through the conversion of passenger vessels to hospital ships had to be made up by the conversion of cargo ships to troop carriers. As the construction of purpose-built hospital ships was out of the question, the Army simply had to take what it could get. Since speed was not a primary concern, the tendency was to obtain and convert, only various older vessels that had proved too slow for convoy use.[1]

By June 1943, ten vessels had been suggested for conversion. Of these, eight were subsequently converted and served as hospital ships; as well as ten other older ships. To these, six newer Liberty ships were added—the subject of this chapter. Collectively, they comprised the Army's 24 hospital ships. Of the eight ships, the oldest that entered service was the *Shamrock* (1907). (Her name and those that follow are the names of the ships during their service as hospital ships, not their preceding ones before conversion.) Next in age, were *Ernest Hinds* and *John L. Clem*, sister ships build in 1918. Almost as old (built in 1920) were the *Chateau Thierry*, *Marigold*, and *St. Mihiel*. *Thistle* (1921) and *Algonquin* (1926) completed the eight vessels originally considered and ultimately converted for use as Army hospital ships.[2]

It is important to note that sequence of selection did not match the order in which ships were eventually converted and placed in service. One striking example is that the ships converted to the Army's first two hospital ships—transport USAT *Acadia* to USAHS *Acadia*, and motor vessel MV *Seminole* to USAHS *Seminole*—were not among the original ten ships considered for conversion. They were part of the other ten older ships.

In 1943, the Army's hospital ship programme was in a state of uncertainty because various candidate vessels could not be utilised, either because they proved unsuitable for the purpose, or because they could not be spared from current assignments. The Army had previously advocated for the conversion of slow-speed passenger vessels and/or newly-built EC2 cargo ships (the so-called Liberty ships). As the Navy vehemently objected to making any Liberty ships available, slow-speed passenger vessels remained the sole source for conversions in the spring and summer of 1943.[3]

However, in autumn 1943 a decision was finally reached to convert six Liberty ships into hospital ships, which were put into service in 1944. Three of these "mercy ships" were named in honour of service members killed earlier in the war:

Blanche F. Sigman: In honour of the 1st Lt. Blanche F. Sigman, the first Army nurse to be killed in action on the Anzio beachhead

Jarrett M. Huddleston: In honour of Col. Jarrett M. Huddleston, Corps Surgeon in the Fifth Army, who was killed in action in Italy early in 1944

John J. Meany: In honour of Maj. John J. Meany, who was killed in action in North Africa in March 1943[4]

Two of the other three ships, *Dogwood* and *Wisteria*, were named for flowers per then Office of the Surgeon General policy; the remaining ship retained her existing name *St. Olaf*, for the Norwegian patron saint.[5]

The Army hospital ships were sailed by Merchant Marine crews. The senior surgeon aboard a hospital ship commanded both the Hospital Ship Complement and the ship itself. On 7 December 1943, total authorised manning for a Hospital Ship Complement was 14 Officers (doctors) – 1 Warrant Officer – 34 Nurses – and 135 Enlisted Men. At the end of June 1945, a total of 481 Commissioned Officers (doctors), 29 Warrant Officers, 1,112 Nurses, and 4,351 Enlisted Men were assigned to Medical Hospital Ship Complements.[6]

REMAINING ARMY HOSPITAL SHIPS

Aside from the Liberty ships, which were originally turned out in 1942-43 as cargo carriers, there were eight remaining additions to the Army hospital fleet in 1944-1945. (Upon reaching its full size of 24 vessels late in the war, the Army's hospital fleet boasted a total patient capacity of more than 14,000.) The final eight vessels in addition to *Acadia* and *Seminole* are identified below by their hospital ship names, and year they were constructed:

- *Larkspur* – 1901
- *Republic* – 1907
- *Charles A. Stafford* – 1918
- *Louis A. Milne* – 1919
- *Ernestine Koranda* – 1919
- *Emily H. M. Weder* – 1920
- *Frances Y. Slanger* – 1927
- *Aleda E. Lutz* – 1931[7]

RETURN TO THE U.S. OF WOUNDED FROM ALLIED INVASIONS OF NORMANDY / SOUTHERN FRANCE

Following their conversion to hospital ships, the six former Liberty ships were employed to return soldiers wounded in the Allied invasions of Normandy and/or southern France home to the United States. The conversion of five of the six hospital ships—*Blanche F. Sigman, Dogwood, Jarrett M. Huddleston, St. Olaf,* and *Wisteria*—took place in New York, following which these vessels sailed for the UK.

In July 1944, the *John J. Meany,* immediately after completion of her conversion at Boston, Massachusetts, left for service in the Mediterranean. She reached Oran in early August, and called at the following ports until leaving Gibraltar for the U.S. in late December: Naples, Palermo, Bari, and Liverno, Italy; and Marseilles, France. The coastal city of Bari (not shown on the map) is located between Naples and Rome. The *Meany* reached Charleston in early January 1945.[8]

Map 21-1

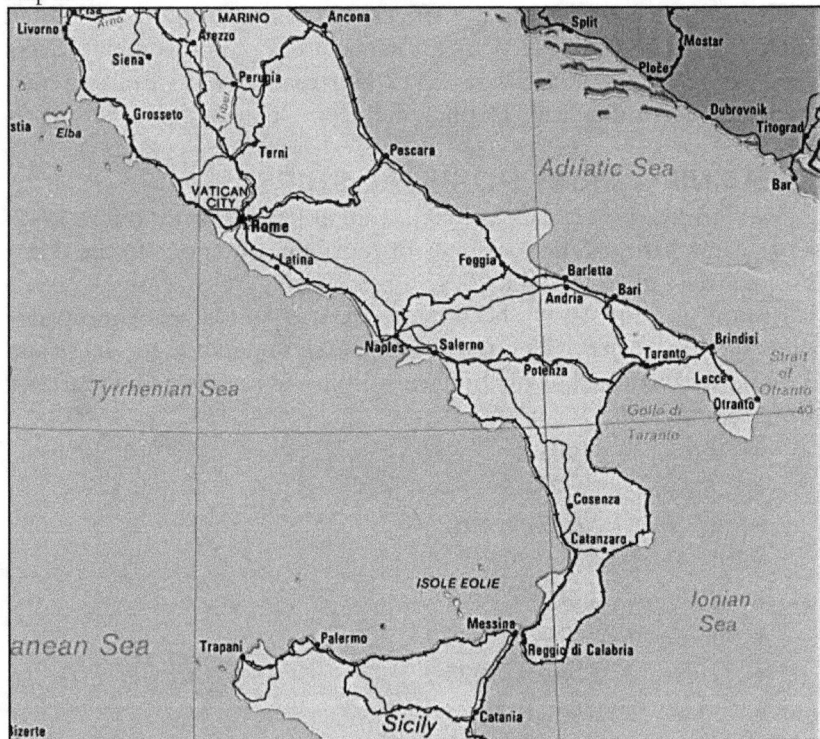

Italy, excluding the most northern and southern regions of the country

The initial voyages of the other five vessels as hospital ships were from New York to English ports and return to New York, before relocation southward to their homeport at Charleston, South Carolina, or arrival directly there.

On her maiden employment as a hospital ship, the *Blanche F. Sigman* sailed from New York in early July 1944 for the Clyde and returned late in the same month. (The Firth of Clyde, estuary of the River Clyde, on the west coast of Scotland is the main entry to Glasgow and several other significant naval and civil ports, plus at the time the site of several major shipyards.) In early August, she left New York and proceeded down the coast to Charleston in South Carolina.[9]

Map 21-2

Charleston is located on the South Carolina coast southeast of Columbia; 630 nautical miles down the U.S. eastern seaboard from New York, NY, and 1,078 from more distant Boston, Massachusetts

Two of the other six ships—*Wisteria* and *Dogwood*—also sailed for England from New York in July 1944. *Wisteria* loaded patients at Liverpool, then made the return Atlantic crossing; reaching Charleston in August. *Dogwood* also returned in August to Charleston, where she was based for the ensuing year. During this period, she made six trips from there to England, usually calling at Avonmouth (port of Bristol) and Liverpool on the River Mersey in England.[10]

Following the end of war in Europe on 7 May 1945, upon the surrender of Germany, *Dogwood* and other U.S. Army hospital ships became available for new duties in the Pacific Theatre of War. In May 1945, *Dogwood* received orders to transit the Panama Canal, continue passage westward to the Philippines, and join the Pacific Fleet. [11]

Map 21-3

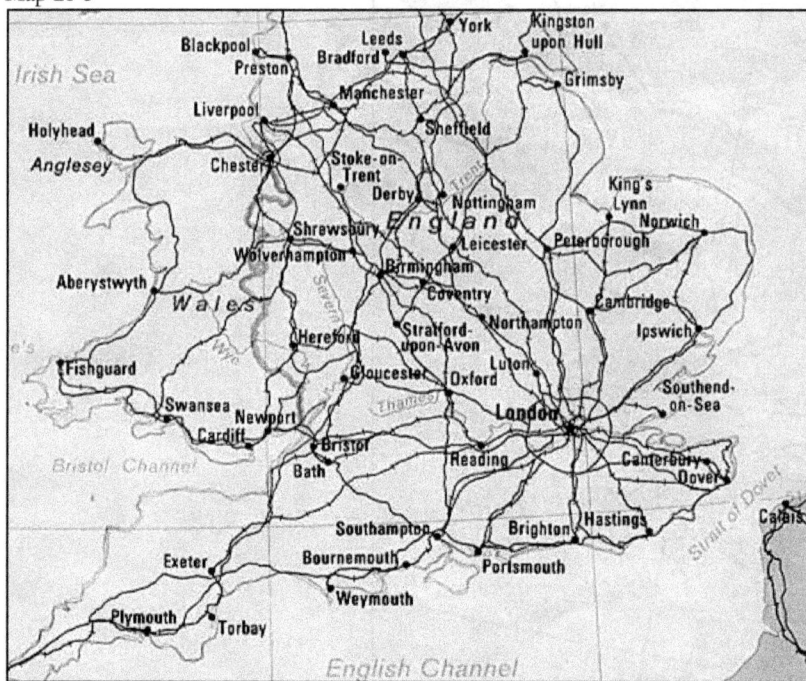

Southern England

In August 1944, the *St. Olaf* sailed from New York to the Clyde and Liverpool, England. She returned from there to Charleston, and in the ensuing eight months, made three trips to Europe (the United Kingdom and Cherbourg). While in shuttle service between Southampton and Cherbourg in early 1945, she suffered minor damage in a collision with another vessel. Returning home from Liverpool in April 1945, *St. Olaf* was ordered to New York for repairs and improvements in her ventilation systems. She left New York in June, bound for Saipan in the Pacific.[12]

USAHS *Jarrett M. Huddleston*, the final of the sixth Liberty ships converted to hospital ships, left New York on her first voyage of mercy on 2 September 1944. She went to Liverpool and Avonmouth to embark wounded soldiers, then returned to Charleston. From there, she made nine voyages to Europe, principally to Avonmouth, although she also went to Falmouth and Cherbourg on some of these trips. While in that area she made thirteen trips across the English Channel between Cherbourg and Southampton from 13 January to 10 March 1945, shuttling sick and wounded.[13]

Map 21-4

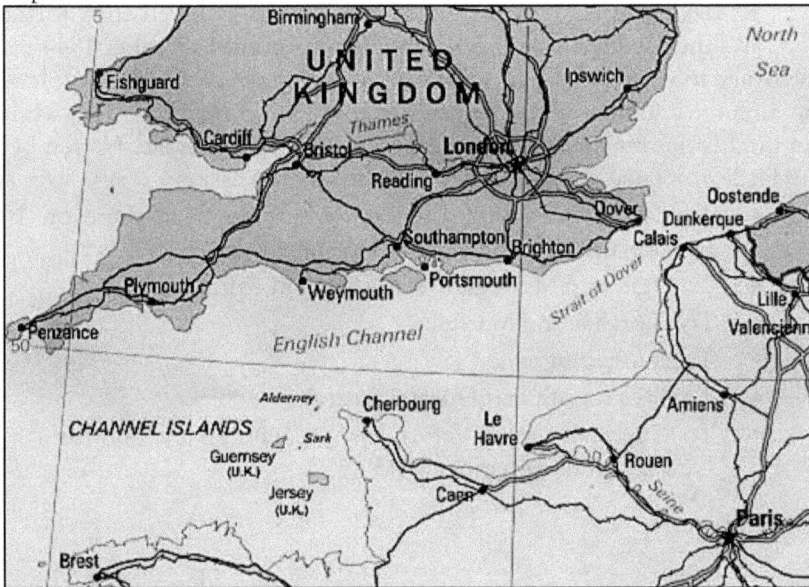

English Channel area

SUBSEQUENT EMPLOYMENT OF *BLANCHE F. SIGMAN* AND *JOHN F. MEANY* AFTER MAIDEN VOYAGES

In late August 1944, *Blanche F. Sigman* voyaged from Charleston to Liverpool. In October, she sailed for Gibraltar, Oran and Liverno; in December she went to the same ports and also to Naples. *Blanche F. Sigman* returned to Charleston in January 1945, and next made the trips as follows through the calendar year 1945:

- To Marseilles and Naples
- To Oran and Marseilles
- To Gibraltar and Marseilles
- To Milford Haven in Wales and Avonmouth, England
- To Cherbourg and Horta, Azores
- To Cherbourg[14]

USAHS *John J. Meany* reached Charleston in early January 1945, returning to the U.S. from her maiden voyage to the Mediterranean. From there, she made a trip to Bermuda and return to Charleston before six voyages to England (Liverpool and Avonmouth) or Cherbourg. She left Southampton, England, in September 1945 and arrived at New York upon return to the United States.[15]

Remaining war employment of the *Wisteria* involved trips to the UK and the Mediterranean. She left Charleston on 1 October 1944 for a voyage to Gibraltar, Oran, and Naples, during which she was involved in a minor collision with a British destroyer HMS *Bazely* (K 311) while in the Mediterranean. After returning to Charleston in late November 1944, *Wisteria* made a series of round trips in succession:

- Two to Avonmouth, England (returning from there on 10 January and 26 February 1945, respectively)
- To Oran and Marseilles (with return in April)
- To Gibraltar and Marseilles
- To Southampton
- To Avonmouth and Horta (Azores)
- To Cherbourg, with return via St. Johns, Newfoundland to New York in September 1945[16]

22

Return Home of POWs
by Hospital Ships

The very fact that SOLOC operated the supply line from the Mediterranean, made it vital to every aspect of French rearmament and rehabilitation. This control of supply lines to the Mediterranean also put SOLOC in a strategic position in the process of repatriation of nationals from Eastern Europe. Territorial contiguity with neutral Switzerland placed SOLOC in position to handle problems incident to exchange of prisoners with Germany.

—History – SOLOC ETOUSA, 9 February 1945. The Southern Line of Communications (SOLOC) was a component of European Theatre of Operations, United States Army. ETOUSA was responsible for directing United Army operations throughout the European theatre of World War II.

Between the completion on 15 September 1944 of Operation DRAGOON (Allied invasion of southern France) and 22 March 1945, when Gen George Patton's U.S. Third Army began crossing the Rhine as part of the Allied invasion of Germany, an exchange of prisoners of war and civilians took place at Marseille, France. This exchange in January 1945 would be the sixth and last of a series of German and American trans-Atlantic exchanges. By then, 2,361 Americans caught behind enemy lines in Europe had been returned from Germany and Italy in exchange for 4,500 Germans and 124 Italians interned in U.S. camps. The most recent exchange with Germany had taken place at Goteborg, Sweden, in September 1944.[1]

On 8 January 1945, SOLOC ETOUSA released a plan for the repatriation of Allied and German sick and wounded prisoners of war and an exchange of civilians through the medium of the Swiss government. This plan resulted from a conference at SOLOC Headquarters on 6 January between the various Allied agencies involved and representatives of the Swiss government. At this meeting, agreements were reached insofar as possible prior to the final agreement between the Swiss and German governments.[2]

The plan called for the exchange to be made in Switzerland during the period 17-31 January 1945. The German prisoners of war (POWs) and civilians were to be assembled by boat at Marseille and then moved into Switzerland on hospital trains provided by the Swiss government. The Allied POWs and civilians were to be moved into Switzerland from Germany and simultaneous exchange effected in the neutral country by the Swiss government. The Allied POWs would then be transported to Marseille, France, by hospital train and embarked aboard ships for their final destination.[3]

Map 22-1

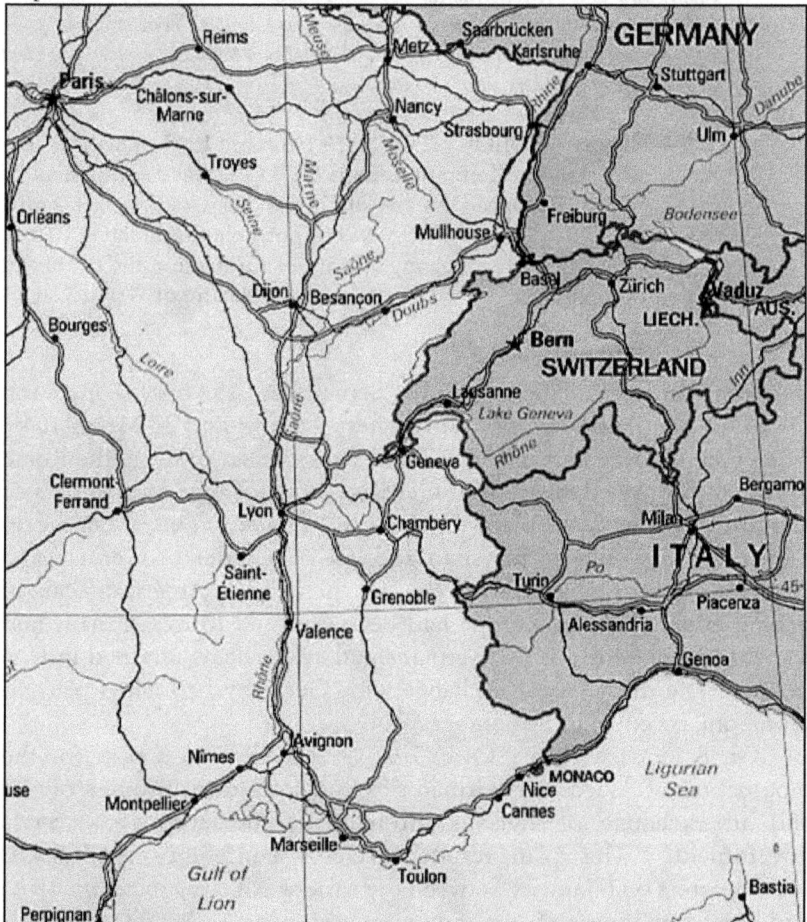

Southwest France and areas of adjacent Italy, Switzerland, and Germany

As noted in the following quoted material, while the exact number of German prisoners of war and civilians to be repatriated by the Allies was specified in the plan, those that were to be repatriated, in turn, by the Germans was not then known:

> The personnel to be repatriated consists of Allied and German prisoners of war that are exchanged on an entire lot basis. The civilians are to be exchanged on a head for head basis. The exchange will be made in two (2) increments. Exchange of the first increment will commence in Switzerland on 17 January 1945. Definite information as to the Allied personnel coming from Germany, both military and civilian, is not yet available.

> The Germans are expected to furnish this information to the Swiss by 10 January 1945, or sooner.... The second increment will be exchanged in Switzerland beginning 25 January 1945. No definite information is available on the Allied personnel coming from Germany. German personnel from British and US sources will consist of 3327 prisoners of war and 1896 civilians.[4]

Information provided in the SOLOC ETOUSA plan regarding the planned prisoner of war exchange, was the best then available, but some details were missing or perhaps incorrect. This is evident in the table, compiled from information available in the plan.

Estimates of Departure Dates from Allied Ports of Origin, and Arrival Dates at Marseille of German Repatriables (2,325)

Departure Date	Place	Ship	Arrival Date	Sources of Prisoners
31 Dec 44	USA	USAHS *Charles Stafford*	10 Jan 45	US 694
13 Jan 45	Naples Algiers	USAHS *Ernest Hinds*	16 Jan 45	US 136 BR 96 FR 32
	Oran	USAHS *Algonquin*	17 Jan 45	US 454
	Delta Base			US 71
11 Jan 45	Mideast	HMHS *Tairea*	15 Jan 45	BR 421
14 Jan 45	Taranto	*Dominion* (believed to be a partial or incorrect vessel name)[5]		Ex-Mideast 205 Ex-Italy 216

To aid in explanation of the table, USAHS *Ernest Hinds*, for example, was scheduled to arrive at Marseille on 16 January 1945, carrying on board a total of 264 German prisoners previously held captive by the United States (136), Britain (96), and France (32).

Four hundred twenty-one prisoners (Ex-Mideast 205, Ex-Italy 216) were to be taken aboard on 14 January at Taranto, in southern Italy, by the vessel *Dominion*. This name is likely incomplete and the ship may have been the SS *Dominion Monarch*.

Delta Base referred to a then existing Allied held area of southern France, whose borders are shown on the cover of the history for this Army unit. Some U.S. Army commands such as the Southern Line of Communications and Delta Base, were quickly created during the war as required, and were disestablished when no longer needed. As shown in the table, 71 of the German prisoners to be exchanged were held in the Delta Base Section of France, and not arriving by ship.

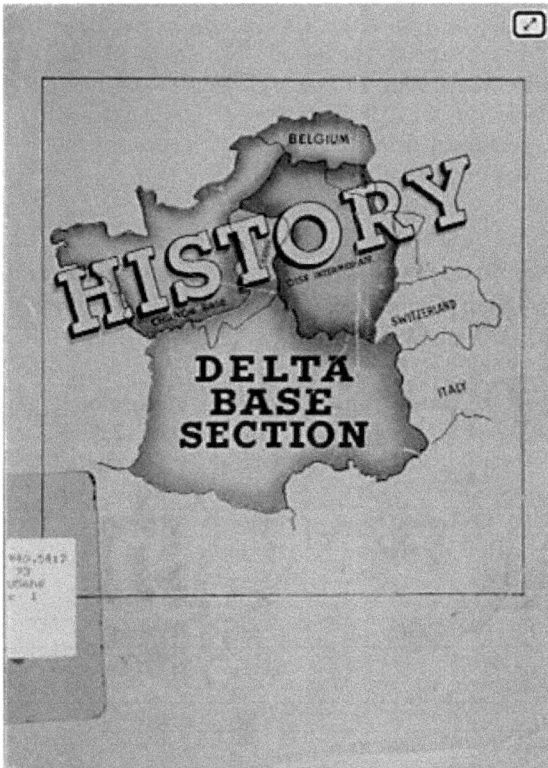

PLANNED MOVEMENT OF PRISONERS BY TRAIN

The Swiss agreed to provide five hospital trains to bring German repatriables from Marseille to Switzerland, and five for the movement of Allied repatriables from Switzerland to Marseille. Each train was to have the capacity of 208 litter cases and 240 ambulatory patients, but these numbers could be varied by configuring a train to conform to the exact numbers of each type of cases. The trains would be completely staffed by Swiss medical personnel, and would not carry any civilians being exchanged.[6]

Because there were no hospital space or staging facilities available in Marseille, it would be necessary to transfer all personnel directly from the trains to ships. As much as practicable, the ships would be berthed at docks having rail facilities immediately adjacent. The Swiss were to request the Germans to load individual trains either with all British or all U.S. personnel so as to simplify the transfer of personnel from train to ship at Marseille.[7]

PLANNED ARRIVAL OF SHIPS TO EMBARK ALLIED FORMER PRISONERS OF WAR

As further noted in the SOLOC ETOUSA plan:

> The estimated arrivals of the vessels carrying the second increment will be furnished in the near future. It is planned to have the *TAIREA* lift British repatriates ex-Marseille for the Mid-East on 2 February 1945. The *LETITIA* will load US and Canadian repatriates, sailing approximately 23 January 1945 for the United States. [Although this may be the plan, she seems to have loaded British like John Doughty, per account at end of chapter.] The *GRIPSHOLM* will lift a mixed load of repatriates and civilians and will sail approximately 31 January 1945 ex-Marseille. The *ARUNDEL CASTLE* will load personnel for United Kingdom, including Australians, to sail approximately 31 January 1945.[8]

The reason for the use of ex-Marseille in the quoted material is unclear. The ships mentioned sailed from Marseille. As shown in the following tables, in the second increment, the motor ship *Gripsholm* and Canadian hospital ship *Letitia* were to collectively bring German prisoners to Marseille. It is unclear what the heading "S/W" and "Ors" used in the plan refers to.

Transport of German Prisoners to Marseille

Ship	Arrive Marseille	S/W	Ors	Total
MS *Gripsholm*	18 Jan 45	300	850	1,150
HMCHS *Letitia*		770		770
				1,920 (total)[9]

The prisoners were to be transferred into three trains for departure for Germany on the dates in the table. MS *Gripsholm* and HMCHS *Letitia*, now empty and waiting, would then embark Allied prisoners, released in turn, for return to their countries. The *Gripsholm* was a Swedish cruise ship chartered by the U.S. government to transport civilians and POWs caught behind enemy lines during World War II. Between 1942 and 1946, she would participate in a dozen exchanges between the U.S. and its wartime enemies; Germany, Italy and Japan.

Swiss Trains Taking German Prisoners from Marseille

Train	Depart Marseille	S/W	Ors	Total
#1	19 Jan 45	274	850	1024
#2	19 Jan 45	448		448
#3	20 Jan 45	448		448

1,920 (total)[10]

RETURN HOME OF STANLEY JOHN DOUGHTY

Photo 22-1

HMCHS hospital ship *Letitia*

12th Royal Lancers
cap badge

Following this introduction, admittedly challenged by a scarcity of information, this chapter closes with a compelling account based on "A Full Life: An Autobiography of Stanley John Doughty 1921-1994." In this work (published by his family) Doughty describes his World War II military service, three years spent in captivity as a German prisoner of war, ultimate freedom, and return to England.

Trooper Stanley John Doughty's unit, the 12th Royal Lancers (part of the First Armoured Division), was engaged in fighting in North Africa on 28 December 1941, when he was captured by German forces. Doughty described his assignment with the 12th Royal Lancers, and the events leading to his imprisonment:

> By nature of our work, reconnaissance, the unit was split into squadrons, that's three cars, most of the time, and very often only one car on its own. The idea was that one should range ahead and

around of the main force, returning to lager at night with our reports. Urgent reports would be radioed in, in Morse code, if possible.

We moved around the desert day by day, backwards, forwards and sideways in pursuit of Jerry who was shortening his lines, sometimes running fast, at other times standing and fighting. Tobruk was the one outpost and port on the Mediterranean which had recently been captured, and from which much of our supplies came.... On the 28th December 1941 we were in the vicinity of Agadabia – a scruffy village surrounding a waterhole, south-west of Tobruk....

We were one car on our own as usual, but with another of the troop within sighting distance, when on putting our Bren [gun] away after a Stuka attack, there was an almighty bang. The interior of the car was full of smoke and both the gunner and I got out as quickly as we could, as the whole thing was going up in flames. Armour burns fast and furious, fed by petrol, rubber and explosives, and I shall remember always the screams of the driver who was trapped and couldn't get out....

Almost at once a German Kuebelwagen came over to investigate, and before I knew what had happened I was sat on top of the car, with Frankie Gowland the gunner, and taken to an interrogation centre in the back of a converted lorry. [The Kuebelwagen was a military light utility vehicle designed by Ferdinand Porsche and built by Volkswagen during World War II.]

The German officer didn't need to ask us anything, we carried our AB64's which gave details of our Army career, and our cars were well known to them since they had captured so many in France. Before long we were, with others, in the back of a British 15cwt truck under guard, off to the town of Misurata [Misrata, Libya] on the coast. There I was singled out from the other prisoners and taken, under escort, to the home jail, a "Beau Geste" sort of place on the waterfront. [*Beau Geste* is an adventure novel by P. C. Wren, which details the adventures of three English brothers who enlist separately in the French Foreign Legion.][11]

This account now skips ahead to near the end of Doughty's POW experience, to he and other German-held prisoners sitting in a draughty hut by a railway station, waiting for what seemed like hours for a train, which would prove to be the first link on their journey to Marseille for repatriation. It took many days to arrive there, for reasons he elaborated on, after first describing the train and its medical capabilities:

...what a train. It seemed to be composed of coaches, not trucks, from all the railways of Europe, in various stages of disrepair. Unfortunately too some of the heating couplings could not be made, so that the travelling turned out to be rather cold. Some prisoners were already on board from other camps and I suppose we made the number up to about 200. Quite a number of stretcher cases, travelling on the floor, and the rest of us, the "walking wounded" as we were termed, on seats which had undoubtedly seen better days.

The atmosphere was electric, only dampened somewhat by our companions saying that they had already been two days on the train. We were soon to discover the truth of this when eventually the train jolted and jarred its way to the Swiss frontier. We collected one more lot of prisoners and were about ten days on the train altogether, being shunted onto sidings in remote places on single track branch lines, as I support the main lines were either bombed or required for military use. We were endlessly counted and checked until even the guards saw the funny side of it all. This travelling menagerie pulled into German Konstance station one morning early amidst great excitement....[12]

There were obvious signs that something was about to happen, as "the platforms were decked with bunting and a military band was tuning up. Within a half hour, an electric train silently pulled in. Doughty described the discharge of its occupants, disappointment with the train's subsequent departure, and finally the arrival of their own train:

From it descended or were carried, the German prisoners which were to be our passport to freedom. They looked well enough, apart from their injuries. The Military Band struck up; there was a speech of welcome; they were all given something in a paper bag, and were then carried away in a fleet of horse-drawn carts. Now, we thought, it would be our turn. We would all be moved into this splendid unreal palace [train], and move quickly through Switzerland.

To our dismay this did not happen. Minutes dragged by. The band packed up and moved off and to our horror the train silently moved off back to Switzerland without us.[13]

The despondency and fear among the men was palpable, but fortunately they were not left long in doubt. Just as silently as the preceding one, a new train came into the same platform, and into it they were helped. Their guards left them, and without a sound the train

moved off across a bridge and pulled up very shortly to a station with a
sign reading CONSTANCE (Suisse).

> The train was immediately invaded by English-speaking nurses, in
> Red Cross uniforms, who helped us out on to the platform which
> was devoid of the public. It had been rigged up with showers and
> toilet facilities, and we were given soap and a towel. Never had a
> wash been so good. About an hour later we were back in the train
> in which we now found English-language papers and magazines on
> little tables between the seats; clean white head rests on the back of
> the seats, and a cloth bag on every seat. These were a gift from
> Switzerland, the bags containing amongst other things, a small block
> of chocolate which they could ill afford, and I thought rather
> incongruously, a small lace-edged handkerchief.... Red Cross
> nurses boarded the train and we were off.... The contrast between
> our past life and this heaven where we were actually treated as
> human beings instead of numbers, left me with an undying
> admiration of the Swiss nation. They could have just passed us
> through, but instead treated us like honoured guests.[14]

It was dark by the time they reached the Swiss/French border, at
which they were transferred to an American staffed hospital train.
Doughty described the train cars as "ex-cattle trucks which had been
properly converted with bunks for every man into which we were
expected to go and stay." Once they were in their bunks, they were each
given a dish of food. Their next stop was arrival at their destination:

> Within hours it seemed, we arrived at Marseille docks, where we
> walked or were carried up the gangplank of HMT [HMCHS] *Letitia*
> – a Canadian hospital ship. Here we stripped, and passed through
> a shower before being on the receiving end of the newly-discovered
> DDT which was squirted everywhere, even into our most intimate
> parts, and re-equipped with Canadian uniforms and underclothing.
> A quick medical check and we were allocated bunks or hammocks,
> by which time the ship was underway.

> The only surprise we met was when we were told that a meal would
> be available as soon as the ship was moving, and naturally we were
> not slow in forming a queue at the mess hall door. Inside we found
> a huge mountain of eggs and bacon, and our ration turned out to be
> half an egg and half a rasher, with one slice of bread. White bread
> too, that hallmark of civilisation, of course. There was a near-riot,
> with the Military Police called in, before a Medical Officer explained
> over the Tannoy that our stomachs would have to be re-educated
> into eating, and that too much would do extreme danger. When he

said that we would be having similar, very small meals every two hours during the day for the whole voyage, most of us saw the sense of it all.[15]

On arrival of HMCHS *Letitia* at Liverpool, the former POWs were immediately taken to an Army hospital by a fleet of coaches and ambulances. Doughty was weighed (just over 6½ stone, 91 pounds), then put through a series of medical tests and X-rays, followed by a visit to the dentist and interrogation officer. He saw no need for the latter requirement, noting, "What they thought we knew about the war after three years in a camp I can't imagine."[16]

At the hospital the men were fed small high-protein meals and after two weeks, split up and sent to hospitals near their homes. Doughty was sent to Brook Hospital (on Shooters Hill in southeast London), where his parents were allowed to visit him for the first time. After a few weeks and yet more examinations, he was allowed home on leave, returning for checkups from time to time, before being discharged from the military and returned to civilian life:

> Eventually, the great day came when I was returned to "Civvy Street." I visited a demob[ilisation] depot somewhere in Town, where I chose the statutory two-piece suit, tie, shoes and a trilby hat; and emerged once again as "Mr Doughty."[17]

Letitia's War and Post-War Duties

Photo 23-1

Hospital ship *Letitia*.
Credit: Canada Department of National Defence/
Library and Archives Canada/ecopy - ID 4232592

Peace finally came to Europe, and VE (Victory in Europe) Day was celebrated on 8 May 1945. All that was left to do then was to repatriate the troops back to Canada in an efficient and orderly manner, and also bring home war brides and war children. It was a long war, and many Canadian servicemen had set down roots and married after 5 long years in England. There were now Canadian dependents overseas who would now have to be moved as well as husbands/fathers back to Canada.[1]

As previously stated, *Lady Nelson* made 30 voyages as a hospital ship in total. Captain Morris Osbourne O'Hara conducted 18, including those to Naples, about which he recalled, "the wounded were loaded on board within the sound of gunfire." The remaining twelve voyages were under O'Hara's predecessor Captain George W. Welch.[2]

Captain William Barclay Armit, RD, RNR, RCNR, Commodore of the Canadian National Steamship Fleet, was appointed master of the *Lady Nelson* on 21 February 1946. Prior to his assignment, *Lady Nelson* had returned to Halifax from Europe on 18 February 1946, where she discharged 486 overseas casualties and notably, three babies. This was her 30th and final trip as a hospital ship, and that of Captain O'Hara as her master. *Lady Nelson* would afterward be engaged in repatriating Canadian war brides and families as part of Operation DADDY under the stewardship of Captain Armit.[3]

HMCHS *LETITIA*

Canada's other hospital ship HMCHS *Letitia* was a much larger vessel capable of carrying 1,000 patients. Prewar, RMS *Letitia* and her sister ship *Athenia* were former Donaldson Atlantic Liner passenger ships. *Athenia* was the first ship torpedoed by the Germans in WWII, suffering the loss of 98 passengers and 19 crew 200 nautical miles northwest of the coast of Northern Island on 3 September 1939. *Letitia* like her former sister was a 525-foot, twin-screw ship. Her six geared steam turbines could propel the 13,475-ton ponderous vessel at a respectable cruising speed of 15 knots.[4]

Letitia's early war service was very similar to that of *Lady Nelson*. Built in 1924-25, she served as a commercial ocean liner until 1939, when she was requisitioned by the Admiralty and commissioned as an armed merchant cruiser. Royal Navy troopship duty followed. In 1944, she was taken over by the Canadian Government and underwent conversion to a hospital ship at Montreal in the Canadian Vickers yard. After 165 days and the expenditure of $3,250,000, she emerged gleaming white with international marking of green horizontal band and red crosses. Said to be "the world's most up-to-the-minute-floating hospital," she was equipped with two operating rooms, labs X-ray machines, dental facilities and even a soda bar for her patients.[5]

From her handover date on 2 November 1944 alongside the Canadian Vickers pier to 11 August 1945, His Majesty's Canadian Hospital Ship *Letitia* completed 16 Atlantic crossing bringing each time on her returns about 700 patients from Britain to Halifax. Her master remaining from before conversion was Captain James Cook, who had previously been master of the ill-fated RMS *Athenia* when torpedoed. Her total crew including her RCAMC complement under Lt. Colonel A.L. Cornish was typically about 410. Most of the deck, engine, cook's and steward department personnel were hired in Britain by the owners, many of whom worked for Donaldson before the war. The medical personnel including Officers, Surgeons, Chaplains, Nursing Sisters, and orderlies totaled about 190. The Nursing Sisters carried were about 35.[6]

HMCHS *Letitia* Ship's Movements during the War in Europe

Departure		Arrival		Days	
Date	Port	Date	Port	At Sea	In Port
10 May 44	Glasgow	25 May 44	Montreal	15	165
6 Nov 44	Montreal	17 Nov 44	Clyde	11	5
22 Nov 44	Clyde	23 Nov 44	Liverpool	1	3
26 Nov 44	Liverpool	8 Dec 44	Halifax	12	5
13 Dec 44	Halifax	22 Dec 44	Falmouth	9	0
22 Dec 44	Falmouth	22 Dec 44	Southampton	0	3
25 Dec 44	Southampton	26 Dec 44	Falmouth	1	0
26 Dec 44	Falmouth	2 Jan 45	Halifax	7	2
4 Jan 45	Halifax	6 Jan 45	New York	2	1
7 Jan 45	New York	18 Jan 45	Gibraltar	11	0
18 Jan 45	Gibraltar	21 Jan 45	Marseille	3	3
24 Jan 45	Marseille	1 Feb 45	Liverpool	8	12
13 Feb 45	Liverpool	22 Feb 45	Halifax	9	6
28 Feb 45	Halifax	10 Mar 45	Liverpool	10	2
12 Mar 45	Liverpool	21 Mar 45	Halifax	9	5
26 Mar 45	Halifax	4 Apr 45	Southampton	9	4
8 Apr 45	Southampton	8 Apr 45	Falmouth	0	0
8 Apr 45	Falmouth	16 Apr 45	Halifax	8	5
21 Apr 45	Halifax	30 Apr 45	Liverpool	9	12
12 May 45	Liverpool	20 May 45	Halifax	8	6
26 May 45	Halifax	4 Jun 45	Southampton	9	3
7 Jun 45	Southampton	16 Jun 45	Halifax	9	6
22 Jun 45	Halifax	30 Jun 45	Avonmouth	8	6
6 Jul 45	Avonmouth	17 Jul 45	Halifax	11	4
21 Jul 45	Halifax	29 Jul 45	Southampton	8	5
3 Aug 45	Southampton	11 Aug 45	Halifax	8	21[7]

As shown in the preceding table summarising *Letitia*'s movements, her voyages as a hospital ship between 6 November 1944 and 11 August 1945 were all between Canadian and the UK ports and return with one exception. Departing Halifax on 4 January 1945, *Letitia* did not follow the usual direct route. Instead, she was diverted to New York where she embarked 850 German civilians and 700 POWs for prisoner exchange. Departing there on 7 January 1945 she proceeded to Marseille, France, via Gibraltar and tied up on a dock where a sealed hospital train containing injured British soldiers had been positioned. As part of the exchange, *Letitia* embarked 6 stretcher cases and 665 walking cases and sailed for Liverpool where she arrived on 1 February. The hospital ship left Liverpool on 13 February, and returned to Halifax on the 22nd with 700 invalided Canadian forces.[8]

Photo 23-2

Nurses and medical staff of *Letitia*, April 1945.
Credit: Canada Dept of National Defence/Library and Archives Canada/PA
- 141658 – ID number 3191562

Following Victory in Europe, the Canadian Government pledged
to have every wounded Canadian serviceman (some 53,000) home by
31 July 1945. To help achieve this, the British hospital ship *Llandoverry
Castle* was assigned to assist *Letitia* and *Lady Nelson*. *Letitia* returned to
Halifax on 16 June with 725 casualties, and a month later on 17 July with
733. As reported the following day in the *Gazette* (Montreal, 18 July),
"As the big ship came into her dock, a band on the pierside struck up
'Roll Out the Barrel' and spontaneous burst of cheering came from the
men lining the rail."[9]

PACIFIC DUTY INTERLUDE

Following surrender of Japan, *Letitia* was diverted to the Pacific via the
Panama Canal to assist in the repatriation of allied troops in cooperation
with US, British, Australian and other forces. Canada's Force C troops
were those sent to aid in the defence of Hong Kong just before the
Japanese invasion of the colony in 1941. Those not killed while resisting
attack became POWs after being forced to surrender. As POWs and

slave labourors, these soldiers had suffered depravity and like many others were in urgent need of repatriation.[10]

Letitia's main mission was repatriation of 1,480 Canadian POWs from the Hong Kong garrison which fell on Christmas Day 1941, and most of whom had been taken to Manila to recover enough to travel. She sailed from Halifax on 1 September 1945, transited the Panama Canal on the 8th and arrived at Pearl Harbor on the 24th. Continuing westward, *Letitia* arrived at Tacloban, Philippines, on 12 October and Manila on 6 November.[11]

Breakdowns and the need for repairs resulted in *Letitia* arriving at Manila much later than scheduled and by then, most of the Canadians had left by other ships and only a few of the original soldiers and a repatriation team that had flown out to assist in the effort remained for her to take to Vancouver. However, her availability was not wasted. Although, she left Manila with but one of the Hong Kong Garrison men, she embarked 695 U.S. Army and Navy wounded personnel, 88 Canadians including Pte. Earl Mossman, last of the Canadian POWs to return, 25 British POWs, French missionaries, British Red Cross nurses and U.S.O. personnel. *Letitia* arrived at Tacoma, Washington, on 5 December 1945, where she disembarked her American personnel before proceeding to Vancouver the following day.[12]

RESUMPTION OF EUROPEAN DUTY

On 14 December, rumor that *Letitia* and *Lady Nelson* might be employed as Canadian war bride ships was confirmed when *Letitia* learned that it was anticipated she would make three round voyages from the U.K. to Halifax every two months, carrying 400 war brides every crossing. To prepare her for this role, *Letitia* was sent to Yarrows Ltd., Esquimalt, for a rush conversion. Following the necessary work, *Letitia* returned to the East Coast where she took up the task along with *Lady Nelson* of bringing war brides and their children to Canada.[13]

Photo 23-3

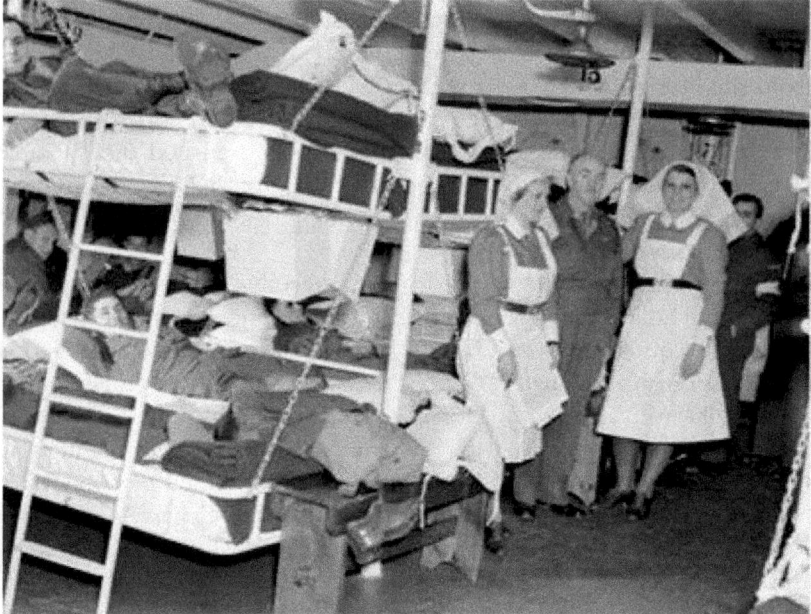

Nursing Sisters and patients in a ward aboard No. 2 Canadian Hospital Ship *Letitia*,
Liverpool England, 21 November 1944.
Credit: Lieut. Charles H. Richard/Canada. Dept of National Defence/Library and
Archives Canada/ PA-128182 – ID number 3398523

Appendix A: WWII Commonwealth Hospital Ships/Carriers

WWII Commonwealth Hospital Ships/Hospital Carriers (49)

HMHS *Aba* Hospital Ship 34	HMHS *Isle of Jersey* Hospital Carrier 3	HMAHS *Oranje* Aus. Hospital Ship 1
HMHS *Amarapoora* Hospital Ship	HMHS *Isle of Thanet* Hospital Carrier 22	HMHS *Oxfordshire* Hospital Ship 6
HMHS *Amsterdam* Hospital Carrier 64	HMHS *Karapara* Hospital Ship 36	HMHS *Paris* Hospital Carrier
HMHS *Atlantis* Hospital Ship 33	HMHS *Karoa* Hospital Ship 60	HMHS *Prague* Hospital Carrier 61
HMHS *Brighton* Hospital Carrier	HMHS *Lady Connaught* Hospital Carrier 55	HMHS *Ramb IV* Hospital Ship
HMHS *Cap St Jacques* Hospital Ship 57	HMCHS *Lady Nelson* Canadian Hospital Ship	HMHS *St Andrew* Hospital Carrier
HMAHS *Centaur* Aus. Hospital Ship 47	HMHS *Leinster* Hospital Carrier	HMHS *St David* Hospital Carrier 27
HMHS *Chantilly* Hospital Ship 63	HMCHS *Letitia* Canadian Hospital Ship	HMHS *St Julien* Hospital Ship
HMHS *Dinard* Hospital Carrier 28	HMHS *Llandovery Castle* Hospital Ship 39	HMHS *Tairea* Hospital Ship 35
HMHS *Dorsetshire* Hospital Ship 23	HMHS *Maid of Kent* Hospital Carrier	HMHS *Talamba* Hospital Ship
HMHS *Duke of Argyll* Hospital Ship 65	RFA *Maine* Hospital ship	HMHS *Takliwa* Hospital Ship
HMHS *Duke of Lancaster* Hospital Ship 56	HMAHS *Manunda* Aus. Hospital Ship 40	HMHS *Tjitjalengaka* Hospital Ship 9
HMHS *Duke of Rothesay* Hospital Carrier 62	HMNZHS *Maunganui* Hospital Ship 42	HMHS *Vasna* Hospital Ship 4
HMHS *El Nil* Hospital Ship 53	HMHS *Naushon* Hospital Carrier 49	HMHS *Vita* Hospital Ship 8
HMHS *Empire Clyde* Hospital Ship 54	HMHS *Newfoundland* Hospital Ship	HMAHS *Wanganella* Aus. Hospital Ship 45
HMHS *Gerusalemme* Hospital Ship 67	HMHS *Ophir* Hospital carrier	HMHS *Worthing* Hospital Carrier 30
HMHS *Isle of Guernsey* Hospital Carrier 26		

Appendix B: U.S. Army Hospital Ships

USAHS	Built	Speed	Capacity	First Voyage / Departure Date	Hospital Unit
Acadia	1932	18	787	New York to N. Africa – 5 Jun 43	204th Med Hosp Ship Co.
Seminole	1907	14	454	New York to N. Africa – 20 Sep 43	
Shamrock	1907	14	543	New York to N. Africa – 4 Sep 43	202nd Med Hosp Ship Co.
Algonquin	1926	15	454	New Orleans to N. Africa – 2 Feb 44	203rd Med Hosp Ship Co.
Thistle	1921	14	455	New York to N. Africa – 8 Apr 44	206th Med Hosp Ship Co.
Chateau Thierry	1920	16	484	Boston to N. Africa – 5 Mar 44	208th Med Hosp Ship Co.
Ernest Hinds	1918	12	288	Charleston to Italy – 14 Jul 44	
Dogwood	1943	11	592	New York to United Kingdom – 21 Jul 44	218th Med Hosp Ship Co.
Larkspur	1901	10	592	Charleston to United Kingdom – 31 Aug 44	209th Med Hosp Ship Co.
St. Mihiel	1920	16	504	Boston to N. Africa – 10 May 44	
Wisteria	1943	11	588	New York to United Kingdom – 16 Jul 44	219th Med Hosp Ship Co.
John L. Clem	1918	12 ½	286	Charleston to N. Africa – 15 Jun 44	
Marigold	1920	12	758	Charleston to Italy – 19 Jul 44	212th Med Hosp Ship Co.
St. Olaf	1942	11	586	New York to United Kingdom – 12 Aug 44	217th Med Hosp Ship Co.
Emily H. M. Weder	1920	13	738	New York to Italy – 12 Jul 44	211th Med Hosp Ship Co.
Jarrett M. Huddleston	1942	11	582	New York to United Kingdom – 2 Sep 44	
John J. Meany	1942	11	582	New York to Italy – 27 Jul 44	
Blanche F. Sigman	1943	11	590	New York to United Kingdom – 7 Jul 44	220th Med Hosp Ship Co.
Charles A. Stafford	1918	16	706	New York to United Kingdom – 21 Sep 44	
Louis A. Milne	1919	12	952	Boston to United Kingdom – 19 Mar 45	200th Med Hosp Ship Co.
Ernestine Koranda	1919	12	722	New York to United Kingdom – 13 Apr 45	
Aleda E. Lutz	1931	16	778	New York to United Kingdom – 18 Apr 45	
Frances Y. Slanger	1927	19	1,628	New York to United Kingdom – 30 Jun 45	235th Med Hosp Ship Co.
Republic	1907	12	1,242	New Orleans to SW Pacific – 4 Sep 45	234th Med Hosp Ship Co.

Appendix C: British Hospital Ships/ Carriers Attacked/Sunk 1 May 1940 - 10 May 1942

Date	Ship	Locality	Attack(s)	Result
1-11 May 40	*Atlantis*	off Norway	bombed on 4 occasions	no damage
18-24 May 40	*Brighton*	Dieppe Harbour	bombed on 3 occasions	sunk
21 May 40	*Maid of Kent*	Dieppe Harbour	bombed	destroyed by fire
24-26 May 40	*St. Andrew*	near Calais	shelled on 2 occasions	no damage
27-31 May 40	*St. Andrew*	near Dunkirk	bombed on 2 occasions	no damage
24-27 May 40	*St. Julien*	Dunkirk; off Calais and in the Downs.	shelled on 2 occasions and bombed once	no damage
29 May 40	*St. Julien*	on passage to Dunkirk.	bombed and machine gunned	slight damage
25 May 40	*St. David*	off Gravelines	shelled from shore batteries	no damage
31 May 40	*St. David*	off Dunkirk	bombed, shelled and machine gunned	no damage
25-26 May 40	*Worthing*	off Calais	shelled on 2 occasions and bombed once	no damage
27 May & 2 Jun 40	*Worthing*	near Dunkirk	bombed and machine gunned on 2 occasions	severe damage
26 May 40	*Isle of Guernsey*	off Calais	shelled from shore batteries	no damage
29 May 40	*Isle of Guernsey*	on passage to Dunkirk.	bombed and machine gunned	slight damage
27 May 40	*Isle of Thanet*	off Calais	shelled from shore batteries	damaged

27 May 40	*Isle of Thanet*	English Channel	bombed and machine gunned	no damage
30 May 40	*Dinard*	off Dunkirk	shelled once and bombed once	slight damage
2-3 Jun 40	*Paris*	on passage to Dunkirk.	bombed on 3 occasions	sunk
31 Jan & 1 Feb 41	*Dorsetshire*	off Sollum	bombed once, machine gunned once	no damage
12 Sep 41	*Dorsetshire*	Port Tewfik	bombed	slight damage
23 Feb 41	*Aba*	Tobruk Harbour	bombed	slight damage
20 April 41	*Aba*	Suda Bay, Crete (at anchor)	bombed	slight damage
16 May 41	*Aba*	Canea, Crete	machine gunned	slight damage
17 May 41	*Aba*	on passage to Haifa	bombed on 2 occasions	damaged
14-21 Apr 41	*Vita*	Tobruk	bombed on 2 occasions	damaged
22 Apr 41	*Vita*	on tow in Mediterranean	bombed	badly damaged
27 Apr 41	*Ramb IV*	near Tobruk	bombed	slight damage
7 Aug 41	*Amra*	Gulf of Suez	torpedoed from aircraft	no damage
5 Sep 41	*Karapara*	Tobruk	bombed	damaged
7 Nov 41	*Llandovery Castle*	Suez (in dock)	bombed	slight damage
7 Dec 41	*Somersetshire*	between Tobruk and Alexandria	dive-bombed and machine-gunned	no damage
30 Jan 42	*Somersetshire*	between Tobruk and Alexandria	Dive-bombed	no damage
21 Feb 42	*Somersetshire*	Tobruk Harbour	bombed	no damage
7 Apr 42	*Somersetshire*	between Alexandria and Tobruk	torpedoed	severe damage
10-12 Feb 42	*Llandovery Castle*	between Alexandria and Tobruk	bombed on 3 occasions	no damage
27 Mar 42	*Llandovery Castle*	Tobruk Harbour	bombed	slight damage
10 May 42	*Ramb IV*	near Alexandria	bombed	sunk

Appendix D: Patients Evacuated from Overseas and Debarked at Army Ports in the U.S.

The below listed totals include U.S. Army, U.S. Navy, and Allied military patients debarked at the Army ports indicated.

Port	1943	1944	1945
Boston, Massachusetts	5,931	8,296	35,383
New York, New York	17,810	39,850	127,748
Baltimore, Maryland	168	1	0
Hampton Roads, Virginia	3,814	12,807	14,800
Charleston, South Carolina	1,128	31,148	41,299
New Orleans, Louisiana	1,261	971	656
Los Angeles, California	1,945	3,528	15,417
San Francisco, California	30,545	45,380	55,789
Seattle, Washington	4,961	2,943	7,925
Total	67,563	144,924	299,017[1]

Bibliography/Notes

Bellafaire, Judith A. *The Army Nurse Corps: A Commemoration of World War II Service*. Washington, D.C.: U.S. Army Center of Military History, 1993.

Bimberg, Edward L. *The Moroccan Goums: Tribal Warriors in a Modern War, 1st Ed.* New York: Praeger, 1999.

Bowden, Jean. *Grey Touched with Scarlet: The War Experiences of the Army Nursing Sisters*. London: Robert Hale, 1959.

Bruhn, David D. *Ingram's Fourth Fleet: U.S. and Royal Navy Operations Against German Runners, Raiders, and Submarines in the South Atlantic in World War II*. Berwyn Heights, MD: Heritage Books, 2017.

—*We Are Sinking, Send Help! The U.S. Navy's Tugs and Salvage Ships in the African, European, and Mediterranean Theaters in World War II*. Berwyn Heights, MD: Heritage Books, 2015.

—*Wooden Ships and Iron Men: The U.S. Navy's Coastal and Motor Minesweepers, 1941-1953*. Westminster, MD: Heritage Books, 2009.

Duckworth, Christian Leslie Dyce and Graham Easton Langmuir. *Railway and other Steamers*. Prescot, Lancashire: T. Stephenson and Sons, 1968.

Feasby, W. R. *Official History of the Canadian Medical Services 1939-1945*. Ottawa: Minister of National Defence, 1956.

Feller, Carolyn M. and Debora R. Cox, *Highlights in the History of the Army Nurse Corps, CMH Pub 85-1*. Washington, D.C.: U.S. Army Center of Military History, 2001.

Garland, Albert N. and Howard McGaw Smyth, *United States Army in World War II Mediterranean Theater of Operations: Sicily and the Surrender of Italy*. Washington, D.C.: Center of Military History, U.S. Army, 1993.

Harris, Roland W. *Troopships of World War II*. Washington, D.C.: Army Transportation Association, 1947.

His Majesty's Stationery Office. *Merchantmen at War: The Official Story of the Merchant Navy, 1939-1944*. London: Ministry of War Transport, 1944.

Larson, Harold. *Army Hospital Ships in World War II*. Washington, D.C.: Office of the Chief of Transportation, Army Service Forces, 1944.

Lott, Arnold S. *Most Dangerous Sea*. Annapolis, MD: U.S. Naval Institute, 1959.

Morison, Samuel Eliot. *Sicily-Salerno-Anzio January 1943-June 1944.*

Edison, NJ: Castle Books, 2001.
—*The Two-Ocean War*. Boston: Little, Brown, 1963.

DEDICATION NOTE:
[1] "18 Reasons Everyone Should Marry a Nurse" by Katy Katz
(https://www.rasmussen.edu/degrees/nursing/blog/reasons-everyone-
should-marry-a-nurse/: accessed 11 February 2024).

PREFACE NOTES:
[1] From Jean Bowden's book *Grey Touched with Scarlet: The War Experiences of the Army Nursing Sisters*, 1959.
[2] "Nursing work and nurses' space in the Second World War:
a gendered construction" (https://manchesteruniversitypress.co.uk/wp-
content/uploads/2018/05/9781526119063_sample.pdf: accessed 4 February 2024).
[3] Ibid.
[4] "QA World War Two Nursing'
(https://www.qaranc.co.uk/qa_world_war_two_nursing.php: accessed 4 February 2024).
[5] Ibid.
[6] "Nursing work and nurses' space in the Second World War:
a gendered construction."
[7] "WWII Hospital Ships" (https://www.med-dept.com/articles/ww2-
hospital-ships/); "Chronology – Utah Beachhead Operations"
(https://www.med-dept.com/unit-histories/chronology-utah-beachhead-
operations/): both accessed 5 February 2024.
[8] "WWII Hospital Ships."
[9] Roland W. Harris, Troopships of World War II (Washington, D.C.: Army Transportation Association, 1947), 327.
[10] "WWII Hospital Ships."
[11] Carolyn M. Feller, Debora R. Cox, Highlights in the History of the Army Nurse Corps, CMH Pub 85-1 (Washington, D.C.: U.S. Army Center of Military History)
https://archive.org/details/HighlightsInTheHistoryOfTheArmyNurseCorps: accessed 14 January 2024).
[12] Ibid.
[13] Ibid.
[14] Ibid.
[15] Ibid.
[16] Ibid.
[17] Ibid.

CHAPTER 2 NOTES:

[1] "How Europe Went To War In 1939"
(https://www.iwm.org.uk/history/how-europe-went-to-war-in-1939:
accessed 13 January 2024).
[2] Ibid.
[3] "How Neutral Norway Fell To The German Blitzkrieg In 1940"
(https://www.iwm.org.uk/history/how-neutral-norway-fell-to-the-german-
blitzkrieg-in-1940: accessed 13 January 2024).
[4] David D. Bruhn, *Ingram's Fourth Fleet: U.S. and Royal Navy Operations Against
German Runners, Raiders, and Submarines in the South Atlantic in World War II*
(Berwyn Heights, MD: Heritage Books, 2017), 21.
[5] "The Altmark Incident – The Royal Navy Freed 299 POW But Caused
Norway To Be Invaded By Nazi Germany World War 2" by Shahan Russell
(https://www.warhistoryonline.com/world-war-ii/the-altmark-incident.html:
accessed 13 January 2024).
[6] Ibid.
[7] "How Neutral Norway Fell To The German Blitzkrieg In 1940."
[8] WW2: The D/S *Queen Maud* is sunk"
(https://web.archive.org/web/20110716085345/http://www.vrakdykking.co
m/maud4.htm: accessed 12 January 2024).
[9] "WW2: The D/S *Queen Maud* is sunk"; "Daily Event for May 1, 2015" by
Michael W. Pocock
(https://www.maritimequest.com/daily_event_archive/2015/05_may/01_dr
onning_maud.htm (both accessed 12 January 2024).
[10] "WW2: The D/S *Queen Maud* is sunk."
[11] "Lehrgeschwader 1" (https://www.asisbiz.com/Luftwaffe/lg1.html:
accessed 13 January 2024); "Daily Event for May 1, 2015" by Michael W.
Pocock.
[12] "Lehrgeschwader 1."
[13] "WW2: The D/S *Queen Maud* is sunk"; "Daily Event for May 1, 2015" by
Michael W. Pocock.
[14] Ibid.

CHAPTER 3 NOTES:

[1] "Hospital Ships" (https://www.qaranc.co.uk/hospitalships.php: accessed
12 January 2024).
[2] "Queen Alexandra's Imperial Military Nursing Service"
(https://www.britishmilitaryhistory.co.uk/docs-services-queen-alexandras-
imperial-military-nursing-service/: accessed 14 January 2024).
[3] Ibid.
[4] Ibid.
[5] Ibid.
[6] "Hospital Ships."
[7] "Scottish Ships" (https://www.clydeships.co.uk/view.php?ref=15582);
"All About Elizabeth Barton, Maid of Kent"

(https://www.englandcast.com/2018/04/elizabeth-barton-maid-of-kent/): both accessed 24 January 2024.
[8] "All About Elizabeth Barton, Maid of Kent."
[9] Ibid.
[10] Ibid.
[11] Ibid.
[12] Ibid.
[13] Ibid.
[14] Ibid.
[15] Ibid.
[16] "German Invasion of Western Europe" (https://encyclopedia.ushmm.org/content/en/article/german-invasion-of-western-europe-may-1940: accessed 15 January 2024).
[17] Ibid.
[18] Ibid.
[19] "Luftwaffe Destruction of the Maid of Kent" by Richard Thwaites (http://www.doverwarmemorialproject.org.uk/Casualties/WWII/Maid%20o f%20Kent.pdf); "TS Maid of Kent (II)" (https://www.doverferryphotosforums.co.uk/ts-maid-of-kent-ii-past-and-present/): both accessed 14 January 2024).
[20] Ut supra.
[21] "TS Maid of Kent (II)"; "Luftwaffe Destruction of the Maid"; "Newhaven" (http://www.dunkirk1940.org/index.php?&p=1_359: accessed 15 January 2024).
[22] "21st May 1940 – Hospital Ship Maid of Kent Bombed, Sunk" (https://www.marpubs.com/21st-may-1940-hospital-ship-maid-of-kent-bombed-sunk/); "TS Brighton (V)" (https://www.doverferryphotosforums.co.uk/ts-brighton-v-past-and-present/): accessed 12 January 2024).
[23] Ut supra.
[24] "21st May 1940 – Hospital Ship Maid of Kent Bombed, Sunk."
[25] "Luftwaffe Destruction of the Maid of Kent."
[26] "TS Maid of Kent (II)."
[27] "Luftwaffe Destruction of the Maid of Kent."
[28] "TS Brighton (V)."

CHAPTER 4 NOTES:

[1] "Nurses at War: At and After Dunkirk" by David Worsfold (https://www.historic-uk.com/HistoryUK/HistoryofBritain/Nurses-At-War-Dunkirk/: accessed 16 January 2024).
[2] Ibid.
[3] Ibid.
[4] "World War II: Admiral Sir Bertram Ramsay" (https://www.thoughtco.com/admiral-sir-bertram-ramsay-2360512: accessed 16 January 2024).

[5] "World War II: Admiral Sir Bertram Ramsay"; "Bertram Ramsay – The Mastermind of the Dunkirk Evacuation Deserves Praise for More than Just Operation Dynamo" (https://militaryhistorynow.com/2020/05/06/bertram-ramsay-the-mastermind-of-the-dunkirk-evacuation-deserves-praise-for-more-than-just-operation-dynamo/: accessed 16 January 2024).

[6] "Dunkirk- The Rescue Fleet and Numbers Rescued" (https://dunkirk1940.org/index.php?&p=1_187: accessed 16 January 2024).

[7] Ibid.

[8] Ibid.

[9] "Winston Churchill's speech before the House of Commons following the disaster at Dunkirk" (https://www.historyonthenet.com/authentichistory/1939-1945/1-war/1-39-41/19400604_Churchill_Address_to_HOC_on_Dunkirk.html: accessed 16 January 2024).

[10] "The miracle of Dunkirk" (https://www.britannica.com/event/Dunkirk-evacuation/The-miracle-of-Dunkirk: accessed 16 January 2024).

[11] "British and Other Navies in World War 2 Day-by-Day" by Don Kindell (https://www.naval-history.net/xDKWW2-4006-19JUN01.htm); "The Miracle of Dunkirk in rare pictures, 1940" (https://rarehistoricalphotos.com/miracle-dunkirk-pictures-1940/): "Wartime Service by Newhaven-Dieppe Steamers As Hospital Ships & Other Duties" by Derek Longly (http://www.ournewhaven.org.uk/page/brighton_vi_and_londres): all accessed 16 January 2024.

[12] "Nurses at War: At and After Dunkirk" (https://www.historic-uk.com/HistoryUK/HistoryofBritain/Nurses-At-War-Dunkirk/: accessed 16 January 2024).

CHAPTER 5 NOTES:

[1] "Italians Bomb Greek Hospital Ship" (https://trove.nla.gov.au/newspaper/article/17726066: accessed 18 January 2024).

[2] "October 28: This is how the Greco-Italian war began in 1940" (https://www.ot.gr/2022/10/28/english-edition/october-28-this-is-how-the-greco-italian-war-began-in-1940/: accessed 18 January 2024).

[3] Ibid.

[4] "About: Greco-Italian War" (https://dbpedia.org/page/Greco-Italian_War: accessed 18 January 2024).

[5] "October 28: This is how the Greco-Italian war began in 1940."

[6] "About: Greco-Italian War"; "British forces arrive in Greece" (https://www.history.com/this-day-in-history/british-forces-arrive-in-greece: accessed 18 January 2024).

[7] "About: Greco-Italian War"; "World War II: Battle of Greece" (https://www.thoughtco.com/world-war-ii-battle-of-greece-2361485: accessed 18 January 2024).

[8] Ut supra.

[9] "Chronology of Events – Crete"
(https://www.britishmilitaryhistory.co.uk/wp-
content/uploads/sites/124/2020/09/Chronology-of-Events-Crete-1.pdf:
accessed 18 January 2024).
[10] Decimation of the Fleet"
(https://greekshippingmiracle.org/en/history/decimation-of-the-fleet-1940-
1945/: accessed 19 January 2024)
[11] "Uncovered! Air attacks against hospital ships: War crimes of Luftwaffe
and Regia Aeronautica in Greece, 1941" by Pierre Kosmidis
(https://greekreporter.com/2013/04/12/sinking-of-hospital-ship-attica-in-
cavo-doro-in-1941/: accessed 18 January 2024).
[12] "Uncovered! Air attacks against hospital ships: War crimes of Luftwaffe
and Regia Aeronautica in Greece, 1941" by Pierre Kosmidis; "Italy's Cant
Z.1007 Alcione was a Three-Engined Wooden Wonder" by Peter Suciu
(https://nationalinterest.org/blog/reboot/italy%E2%80%99s-cant-z1007-
alcione-was-three-engined-wooden-wonder-189909: accessed 18 January
2024).
[13] "Uncovered! Air attacks against hospital ships: War crimes of Luftwaffe
and Regia Aeronautica in Greece, 1941" by Pierre Kosmidis.
[14] "Scottish Built Ships"
(https://www.clydeships.co.uk/view.php?ref=23790); "Sinking of Hospital
Ship Attica in Cavo D'oro in 1941" by Stella Tsolakidou
(https://greekreporter.com/2013/04/12/sinking-of-hospital-ship-attica-in-
cavo-doro-in-1941/): both accessed 18 January 2024.
[15] "Stuka German aircraft"
(https://www.britannica.com/technology/Stuka); "The Sirens of Death – 11
Amazing Facts About the Ju 87 Stuka"
(https://militaryhistorynow.com/2015/06/04/screaming-death-10-amazing-
facts-about-the-ju-87-stuka/): accessed 19 January 2024).
[16] Ut supra.
[17] "Sinking of Hospital Ship Attica in Cavo D'oro in 1941."
[18] Ibid.
[19] "Balkans Campaign 28 Oct 1940 - 1 Jun 1941" by C. Peter Chen
(https://ww2db.com/battle_spec.php?battle_id=36: accessed 19 January
2024).
[20] "Uncovered! Air attacks against hospital ships: War crimes of Luftwaffe
and Regia Aeronautica in Greece, 1941"; "SS *Esperos* (+1941)"
(https://www.wrecksite.eu/wreck.aspx?139519); "British and Other Navies
in World War 2 Day-by-Day" (by Don Kindell (https://www.naval-
history.net/xDKWW2-4104-31APR02.htm): all accessed 18 January 2024).
[21] "British and Other Navies in World War 2 Day-by-Day"
by Don Kindell (https://www.naval-history.net/xDKWW2-4104-
31APR02.htm: accessed 18 January 2024).
[22] "Balkans Campaign 28 Oct 1940 - 1 Jun 1941."

[22] "British and Other Navies in World War 2 Day-by-Day"; "Uncovered!
Air attacks against hospital ships: War crimes of Luftwaffe and Regia
Aeronautica in Greece, 1941."
[23] "Uncovered! Air attacks against hospital ships: War crimes of Luftwaffe
and Regia Aeronautica in Greece, 1941."
[24] "Balkans Campaign 28 Oct 1940 - 1 Jun 1941."
[25] Ibid.
[26] "British and Other Navies in World War 2 Day-by-Day."
[27] "Sinking of the Hellas 24 April 1941"
(https://nzhistory.govt.nz/page/hellas-
sinking#:~:text=When%20the%20Hellas%2C%20a%20large%20steam%20y
acht%2C%20arrived,and%20an%20Australian%20hospital%20were%20sent
%20on%20board: accessed 19 April 2024).
[28] Ibid.
[29] Ibid.

CHAPTER 6 NOTES:
[1] "31 Willful, Brutal Nazi Attacks On Hospital Ships"
The Northern Advocate, 4 July 1941
(https://paperspast.natlib.govt.nz/newspapers/NA19410704.2.44: accessed
20 January 2024).
[2] "31 Willful, Brutal Nazi Attacks On Hospital Ships"; "Allied Hospital
Ships (German And Italian Sinkings) Volume 389: debated on Thursday 27
May 1943" (https://hansard.parliament.uk/Commons/1943-05-
27/debates/8b2710b7-d858-430b-a243-
8efcc96e34cf/AlliedHospitalShips(GermanAndItalianSinkings): both accessed
20 January 2024.
[3] "Allied Hospital Ships (German And Italian Sinkings) Volume 389:
debated on Thursday 27 May 1943."
[4] "The Red Sea Flotilla" by Giulio Poggiaroni
(https://comandosupremo.com/the-red-sea-flotilla/: accessed 21 January
2024).
[5] "Italian Hospital Ship Ramb Iv" (https://academic-
accelerator.com/encyclopedia/italian-hospital-ship-ramb-iv); "Dictionary of
Egyptian Shipwrecks – R"
(https://web.archive.org/web/20111122211841/http://www.shipwrecksofe
gypt.com/images/shippages/rambiv.html) both accessed 21 January 2024).
[6] "Dictionary of Egyptian Shipwrecks – R"
(https://www.deeplens.com/dictionary-of-egyptian-shipwrecks-r/);
"Carpenter Erich Karo"
(https://www.veterans.gc.ca/eng/remembrance/memorials/canadian-virtual-
war-memorial/detail/2558584): both accessed 20 January 2024.

CHAPTER 7 NOTES:

[1] Bruhn, *We Are Sinking, Send Help! The U.S. Navy's Tugs and Salvage Ships in the African, European, and Mediterranean Theaters in World War II* (Berwyn Heights, MD: Heritage Books, 2015), 1.

[2] Ibid, 1.

[3] Ibid, 2.

[4] Ibid.

[5] Ibid.

[6] Ibid, 3-4.

[7] Commander Task Force Thirty-four, Torch Operation, preliminary report of, 28 November 1942.

[8] Ibid.

[9] *Hamilton* War Diary, November 1942; Arnold S. Lott, *Most Dangerous Sea* (Annapolis, MD: U.S. Naval Institute, 1959), 107; Bruhn, *We Are Sinking, Send Help!*, 8; ONI Combat Narratives, The Landings in North Africa 1942, 1944.

[10] "48th Surgical/128th Evacuation Hospital Unit History" (https://www.med-dept.com/unit-histories/48th-surgical-128th-evacuation-hospital/: accessed 25 January 2024).

[11] 48th Surgical/128th Evacuation Hospital Unit History"; USS *Wakefield*, *DANFS*.

[12] "48th Surgical/128th Evacuation Hospital Unit History."

[13] Transport Division Eleven, Amphibious Force, U.S. Atlantic Fleet War Diary, November 1942.

[14] Judith A. Bellafaire, *The Army Nurse Corps: A Commemoration of World War II Service* (Washington, D.C.: U.S. Army Center of Military History, 1993), 9-10.

[15] Ibid.

[16] Ibid.

[17] Ibid, 11.

[18] Ibid, 12.

[19] Ibid.

[20] Ibid.

[21] Ibid.

[22] Ibid, 12-13.

[23] Ibid, 13.

[24] Ibid.

CHAPTER 8 NOTES:

[1] Harold Larson, *Army Hospital Ships in World War II* (Washington, D.C.: Office of the Chief of Transportation, Army Service Forces, 1944), 17 (https://archive.org/details/ArmyHospitalShipsInWorldWarIi: accessed 9 January 2024).

[2] Larson, *Army Hospital Ships in World War II*, 17-18.

[3] Larson, *Army Hospital Ships in World War II*, 20; "Eisenhower in 1942" (https://www.nps.gov/articles/000/eisenhower-in-1942.htm: accessed 26 January 2024).

[4] Larson, *Army Hospital Ships in World War II*, 22-23.
[5] Ibid, 20-23.
[6] Charles, *Troopships of World War II*, 328.
[7] Larson, *Army Hospital Ships in World War II*, 23.
[8] Ibid, 23-24.
[9] Ibid, 24.
[10] Larson, *Army Hospital Ships in World War II*, 24; "Veteran's Testimony –
Martin Lipschultz 204th Medical Hospital Ship Company - USAHS *Acadia*"
(https://www.med-dept.com/veterans-testimonies/veterans-testimony-
martin-lipschultz/: accessed 26 January 2024).
[11] "Veteran's Testimony – Martin Lipschultz 204th Medical Hospital Ship
Company - USAHS *Acadia*."
[12] "Veteran's Testimony – Martin Lipschultz 204th Medical Hospital Ship
Company - USAHS *Acadia*"; Larson, Army Hospital Ships in World War II,
24.
[13] Larson, *Army Hospital Ships in World War II*, 24-25.
[14] Ibid, 24-25.
[15] Ibid, 26-27.
[16] Larson, *Army Hospital Ships in World War II*, 26-27; "Veteran's Testimony
– Martin Lipschultz 204th Medical Hospital Ship Company - USAHS *Acadia*";
"Timeline USAHS *Acadia* – Lt. Colonel Thomas B. Protzman"
(https://www.med-dept.com/veterans-testimonies/timeline-usahs-acadia-lt-
colonel-thomas-b-protzman/: accessed 26 January 2024).
[17] "Veteran's Testimony – Martin Lipschultz 204th Medical Hospital Ship
Company - USAHS *Acadia*"; "Timeline USAHS *Acadia* – Lt. Colonel Thomas
B. Protzman."
[18] "Veteran's Testimony – Martin Lipschultz 204th Medical Hospital Ship
Company - USAHS *Acadia*."
[19] "Timeline USAHS *Acadia* – Lt. Colonel Thomas B. Protzman."
[20] Ibid.
[21] Ibid.
[22] Ibid.
[23] Ibid.
[24] Larson, *Army Hospital Ships in World War II*, 29.
[25] "Timeline USAHS *Acadia* – Lt. Colonel Thomas B. Protzman."
[26] Larson, *Army Hospital Ships in World War II*, 29.
[27] Larson, *Army Hospital Ships in World War II*, 28; Veteran's Testimony –
Martin Lipschultz 204th Medical Hospital Ship Company - USAHS *Acadia*."
[28] "Timeline USAHS *Acadia* – Lt. Colonel Thomas B. Protzman."
[29] Ibid.
[30] Ibid.
[31] Ibid.

CHAPTER 9 NOTES:

[1] "Timeline USAHS *Acadia* – Lt. Colonel Thomas B. Protzman."
[2] Ibid.

[3] Ibid.
[4] Ibid.
[5] Ibid.
[6] Ibid
[7] Ibid.
[8] Ibid.
[9] Ibid.
[10] Ibid.
[11] Dwight Messimer correspondence of 31 July 2018.
[12] Ibid
[13] Ibid.
[14] Bruhn, *We Are Sinking, Send Help!*, 53.
[15] Ibid, 53-54.
[16] Ibid, 54.
[17] Ibid, 59.
[18] "Timeline USAHS *Acadia* – Lt. Colonel Thomas B. Protzman."
[19] Ibid.
[20] Ibid.
[21] Ibid.
[22] Ibid.
[23] Ibid.
[24] Ibid.
[25] Ibid.
[26] Ibid.
[27] Ibid.
[28] Ibid.
[29] Ibid.
[30] Ibid.
[31] Ibid.
[32] Ibid.
[33] Ibid.
[34] Ibid.
[35] Ibid.
[36] Ibid.
[37] "Timeline USAHS *Acadia* – Lt. Colonel Thomas B. Protzman"; Edward L. Bimberg, *The Moroccan Goums: Tribal Warriors in a Modern War, First Edition* (New York: Praeger, 1999).
[38] "Timeline USAHS *Acadia* – Lt. Colonel Thomas B. Protzman."
[39] Ibid.
[40] Ibid.
[41] Ibid.
[42] Ibid.
[43] Ibid.
[44] Ibid.
[45] Ibid.
[46] Ibid.

[47] Ibid.
[48] Ibid.
[49] Ibid.
[50] Ibid.
[51] Ibid.
[52] Ibid.
[53] Ibid.
[54] Bruhn, *We Are Sinking, Send Help!*, 81.
[55] Ibid, 81-82
[56] Ibid.
[57] Ibid, 82.

CHAPTER 10 NOTES:

[1] *"Talamba"* (https://www.tynebuiltships.co.uk/T-Ships/talamba1924.html: accessed 22 January 2024).
[2] *"Talamba"*; "1937 SS *Talamba* off Devil's Peak"
(https://gwulo.com/media/15100): accessed 22 January 2024.
[3] Ut supra.
[4] "Sunk By Enemy Action – SS *Talamba*"
(https://tyneareasc.org.uk/2012/10/sunk-by-enemy-action-s-s-talamba/: accessed 22 January 2024).
[5] Ibid.
[6] Ibid.
[7] Ibid.
[8] Ibid.
[9] Ibid.
[10] Ibid.
[11] Ibid.
[12] CincMed War Diary, 1 July 1943-31 August 1943.

CHAPTER 11 NOTES

[1] ComEighthFlt (CWNTF), The Italian Campaign, 11 January 1945.
[2] ComEighthFlt (CWNTF), The Italian Campaign; Samuel Eliot Morison, *The Two-Ocean War* (Boston: Little, Brown, 1963), 350.
[3] ComEighthFlt (CWNTF), The Italian Campaign.
[4] Morison, *The Two-Ocean War*, p. 350; Robert M. Citino, "Avalanche: How Both Sides Lost at Salerno" (http://www.historynet.com/avalanche-how-both-sides-lost-at-salerno.htm: accessed 9 December 2013).
[5] Morison, *The Two-Ocean War*, p. 353.
[6] "RN Beach Commandos and Operation Avalanche The Salerno Landings September 9th 1943"
(http://www.relaysystem.co.uk/RNBC_1943Salerno.pdf: accessed 23 December 2013); J. D. Lock, "World War II – North Africa/Europe" (http://www.armyranger.com/index.php/history/modern-era: accessed 25 December 2013).

[7] "Patton's Career A Brilliant One," obituary in the New York Times, 22 December 1945; Citino, "Avalanche: How Both Sides Lost at Salerno".
[8] Commander U.S. Naval Forces, Northwest African Waters, War Diary; forwarding of, 6 April 1944.
[9] Ibid.
[10] Ibid.
[11] ComEighthFlt (CWNTF), The Italian Campaign.
[12] ComEighthFlt (CWNTF), The Italian Campaign; ComEighthFlt, Action Reports for the Salerno Operation, 7 March 1945.
[13] ComEighthFlt (CWNTF), The Italian Campaign.
[14] Ibid.
[15] CTU 85.1.7, Report of Salvage Operation on "UNCLE" Beaches during Avalanche Operation.
[16] Morison, *The Two-Ocean War*, 356.
[17] Ibid.
[18] Ibid, 356-357.
[19] Ibid, 357.
[20] "Timeline USAHS *Acadia* – Lt. Colonel Thomas B. Protzman."
[21] Ibid.
[22] Ibid.
[23] Ibid.
[24] Ibid.
[25] Ibid.
[26] Ibid.
[27] Ibid.
[28] Ibid.
[29] Ibid.
[30] Ibid.

CHAPTER 12 NOTES:

[1] Commander United States Eighth Fleet, The Italian Campaign, 11 January 1945.
[2] Lloyd's Register of Shipping (https://plimsoll.southampton.gov.uk/shipdata/pdfs/35/35b0594.pdf: accessed 21 January 2024).
[3] "The Sinking of HMHS *Newfoundland*" (https://www.bbc.co.uk/history/ww2peopleswar/stories/71/a3412171.shtml: accessed 21 January 2024).
[4] "The Sinking of HMHS *Newfoundland*"; "HMHS *Newfoundland*" (https://britisharmynurses.com/hmhs-newfoundland/); "An Army Nurse Describes a Deadly Attack on a Hospital Ship" by Andrew Carroll (http://www.historynet.com/an-army-nurse-describes-a-deadly-attack-on-a-hospital-ship.htm): all accessed 21 January 2024.
[5] "HMHS *Newfoundland*."
[6] "An Army Nurse Describes a Deadly Attack on a Hospital Ship."
[7] "The Sinking of HMHS *Newfoundland*."

[8] "The Sinking of HMHS *Newfoundland*"; "HMHS *Newfoundland*."
[9] "The Sinking of HMHS *Newfoundland*"; Commander United States Eighth Fleet, The Italian Campaign, 11 January 1945.
[10] Commander Destroyer Squadron Seven War Diary, September 1943.
[11] Ibid.
[12] Commander Destroyer Squadron Seven War Diary, September 1943; Bruhn, *We Are Sinking, Send Help! The U.S. Navy's Tugs and Salvage Ships in the African, European, and Mediterranean Theaters in World War II*, 104.
[13] Bruhn, *We Are Sinking, Send Help! The U.S. Navy's Tugs and Salvage Ships in the African, European, and Mediterranean Theaters in World War II*, 104.
[14] Ibid.
[15] Ibid.
[16] "HMHS *Newfoundland*."
[17] Bellafaire, *The Army Nurse Corps: A Commemoration of World War II Service*.

CHAPTER 13 NOTES:

[1] "USAHS *Shamrock*" (https://www.navsource.org/archives/09/12/1203.htm: accessed 1 February 2024).
[2] *Shamrock, DANFS*.
[3] Charles, *Troopships of World War II*, 5.
[4] Larson, *Army Hospital Ships*, 41.
[5] Ibid, 41-42.
[6] Ibid, 43, 47.
[7] Ibid, 43.
[8] Ibid, 47.
[9] "Why a World War II ship was named after a West Rockhill nurse" by Scott Bomboy (https://preservingperkasie.com/2019/06/06/why-a-world-war-ii-ship-was-named-after-a-west-rockhill-nurse/: accessed 1 February 2024).
[10] Samuel Eliot Morison, *Sicily-Salerno-Anzio January 1943-June 1944* (Edison, NJ: Castle Books, 2001), 303.
[11] Ibid, 304.
[12] "Timeline USAHS *Acadia* – Lt. Colonel Thomas B. Protzman."
[13] USS *Bernadou* War Diary, September 1943; Commander, United States Eighth Fleet, The Italian Campaign, 11 January 1945.
[14] "Timeline USAHS *Acadia* – Lt. Colonel Thomas B. Protzman."
[15] Ibid.
[16] Commander, Mine Division Seventeen, Action Report, USS *Skill*, Submission of, 1 October 1943; USS *Skill, DANFS*; "USS *Skill* (AM 115)" (https://uboat.net/allies/merchants/ship/3085.html: accessed 11 December 2022).
[17] Commander, Mine Division Seventeen, Action Report, USS *Skill*, Submission of, 1 October 1943; "Timeline USAHS *Acadia* – Lt. Colonel Thomas B. Protzman."
[18] "Timeline USAHS *Acadia* – Lt. Colonel Thomas B. Protzman."

[19] Ibid.
[20] Ibid.
[21] Ibid.
[22] Ibid.
[23] Ibid.
[24] Ibid.
[25] Ibid.
[26] ComEighthFlt (CWNTF), The Italian Campaign.
[27] Ibid.
[28] ComEighthFlt (CWNTF), The Italian Campaign.
[29] Albert N. Garland and Howard McGaw Smyth, *United States Army in World War II Mediterranean Theater of Operations: Sicily and the Surrender of Italy* (Washington, D.C.: Center of Military History, U.S. Army, 1993), p. 552.
[30] Garland and Smyth, *United States Army in World War II Mediterranean Theater of Operations: Sicily and the Surrender of Italy*, p. 552; Citino, "Avalanche: How Both Sides Lost at Salerno."
[31] Ibid.

CHAPTER 14 NOTES:

[1] "The Lady Boats" (https://legionmagazine.com/the-lady-boats/: accessed 8 February 2024).
[2] *Lady Nelson* (https://uboat.net/allies/merchants/ship/1418.html; "*Umtata*" (https://uboat.net/allies/merchants/ship/1419.html); "The Life and Times of Captain Morris O'Hara (Isaacs Harbour, NS), Master of *Lady Nelson* during the Second World War" (http://www.forposterityssake.ca/RCN-DOCS/Capt-Morris-OHara.pdf): all accessed 8 February 2024.
[3] "*Lady Nelson*"; "The Life and Times of Captain Morris O'Hara (Isaacs Harbour, NS), Master of *Lady Nelson* during the Second World War."
[4] "The Lady Boats"; "The Life and Times of Captain Morris O'Hara (Isaacs Harbour, NS), Master of *Lady Nelson* during the Second World War"; "Taylor was chief medical officer of 'mercy ship' HMCS *Lady Nelson* by David Yates For the Signal Star (https://www.clintonnewsrecord.com/opinion/columnists/taylor-was-chief-medical-officer-of-mercy-ship-hmcs-lady-nelson: accessed 18 February 2024).
[5] W. R. Feasby, *Official History of the Canadian Medical Services 1939-1945* (Ottawa: Minister of National Defence, 1956), 128 (https://www.canada.ca/content/dam/themes/defence/caf/militaryhistory/dhh/official/book-1956-medical-services-1-en.pdf: accessed 18 February 2024).
[6] "The Life and Times of Captain Morris O'Hara (Isaacs Harbour, NS), Master of Lady *Nelson* during the Second World War."
[7] "Canadian Nursing Sisters, 1939-1945" (https://www.silverhawkauthor.com/post/women-in-the-canadian-forces-nursing-sisters); "The Nursing Sisters of Canada" (https://www.veterans.gc.ca/eng/remembrance/those-who-served/women-

veterans/nursing-sisters#Introduction); "Canadian Nurses in Military Service 1895 – Present" (https://www.royalcdnmedicalsvc.ca/wp-content/uploads/2019/11/Canadian-Nurses-Album.pdf): all accessed 8 February 2024.

[8] "Canadian Nursing Sisters, 1939-1945"; "The Nursing Sisters of Canada" (https://www.veterans.gc.ca/eng/remembrance/those-who-served/women-veterans/nursing-sisters#Introduction): both accessed 8 February 2024.

[9] Ut supra.

[10] "The Nursing Sisters of Canada"; "Canadian Nurses in Military Service 1895 – Present."

[11] "The Nursing Sisters of Canada"; "SS *Santa Elena*" (https://uscs.org/ss-santa-elena/); SS Santa Elena" (https://www.wrecksite.eu/wreck.aspx?96349): both accessed 8 February 2024.

[12] "Sub-Lieutenant Agnes Wilkie" (https://www.communitystories.ca/v2/tragic-sinking-ss-caribou_naufrage-tragique/gallery/sub-lieutenant-agnes-wilkie/: accessed 9 February 2024).

[13] "Sinking of the SS *Caribou*" (https://thecanadianencyclopedia.ca/en/article/sinking-of-the-ss-caribou); "*Caribou*" (https://uboat.net/allies/merchants/ship/2270.html): both accessed 9 February 2024.

[14] Ut supra.

[15] "The Nursing Sisters of Canada."

[16] Ibid.

[17] "Angels of Mercy": Canada's Nursing Sisters in World War I and II (https://digitalcollections.mcmaster.ca/pw20c/case-study/angels-mercy-canadas-nursing-sisters-world-war-i-and-ii: accessed 8 February 2024).

CHAPTER 15 NOTES:

[1] Larson, *Army Hospital Ships in World War II*, 55.

[2] Ibid, 57.

[3] Ibid.

[4] Ibid, 59.

[5] Ibid, 69.

[6] Ibid, 63.

[7] Ibid, 69.

[8] Ibid, 59-60.

[9] Ibid.

CHAPTER 16 NOTES:

[1] Bellafaire, *The Army Nurse Corps: A Commemoration of World War II Service*.

[2] Ibid.

[3] "Great Western Railway" (https://www.simplonpc.co.uk/GWR1.html: accessed 23 January 2024); Duckworth, Christian Leslie Dyce and Graham Easton Langmuir, *Railway and other Steamers* (Prescot, Lancashire: T.

Stephenson and Sons, 1968); "New Railway Steamer," *Hartlepool Mail*, Hartlepool, 10 December 1931.

[4] Commanding Officer, USS *Sway*, Action Report – Anzio Landing Operation, 4 March 1944.

[5] Ibid.

[6] Ibid.

[7] Ibid.

[8] Admiralty War Diaries, 4 February 1944.

[9] Ibid.

[10] Ibid.

[11] "hospital ship SS St David" (https://www.shipsnostalgia.com/threads/hospital-ship-ss-st-david.40189/: accessed 23 January 2024).

[12] Ibid.

[13] Ibid.

[14] Ibid.

[15] Ibid.

[16] Ibid.

[17] "Hospital Ships" (https://www.qaranc.co.uk/hospitalships.php: accessed 24 January 2024).

CHAPTER 17 NOTES:

[1] Bruhn, *Wooden Ships and Iron Men: The U.S. Navy's Coastal and Motor Minesweepers, 1941-1953* (Westminster, MD: Heritage Books, 2009), 95.

[2] Fact Sheet: Normandy Landings (https://obamawhitehouse.archives.gov/the-press-office/2014/06/06/fact-sheet-normandy-landings: accessed 18 August 2018).

[3] Commander Assault Force "O", Action Report – Assault on Vierville-Colleville Sector, Coast of Normandy, 27 July 1944.

[4] Ibid.

[5] Historical Reports and Monographs - Staff Section Reports File Number 581-12, Surgeon General, Medical History, ETO Vol XII Conflict Period World War II.

[6] Report by The Allied Naval Commander-in-Chief Expeditionary Force on Operation Neptune, Volume I, 1944; His Majesty's Stationery Office, *Merchantmen at War: The Official Story of the Merchant Navy, 1939-1944* (London: Ministry of War Transport, 1944).

[7] Report of Operations Hq First U.S. Army for period 20 October 1943 – 1 August 1944.

[8] Report by The Allied Naval Commander-in-Chief Expeditionary Force on Operation Neptune, Volume I, 1944.

[9] His Majesty's Stationery Office, *Merchantmen at War: The Official Story of the Merchant Navy, 1939-1944*.

[10] Report of Operations Hq First U.S. Army for period 20 October 1943 – 1 August 1944.

[11] Commander Assault Force "O", Action Report – Assault on Vierville-Colleville Sector, Coast of Normandy, 27 July 1944.
[12] Commander Assault Force "O", Action Report – Assault on Vierville-Colleville Sector, Coast of Normandy, 27 July 1944.

CHAPTER 18 NOTES:
[1] "Canada in the Second World War" (https://www.junobeach.org/canada-in-wwii/articles/the-army-medical-organization/: accessed 9 February 2024).
[2] Ibid.
[3] Ibid.
[4] Ibid.
[5] Ibid.
[6] Ibid.
[7] Ibid.
[8] Ibid.
[9] Ibid.
[10] "Canada in the Second World War" (https://www.junobeach.org/canada-in-wwii/articles/the-army-medical-organization/: accessed 9 February 2024).
[11] Ibid.
[12] Ibid.

CHAPTER 19 NOTES:
[1] "*Amsterdam* 1930 HMHS - Hospital Carrier" (https://www.clydemaritimeforums.co.uk/amsterdam-1930-hmhs-hospital-carrier-t7462.html: accessed 24 January 2924).
[2] "*Amsterdam* 1930 HMHS - Hospital Carrier"; "The sinking of the S.S. *Amsterdam* [Hospital ship] 1944" by Patrick Manning (https://www.bbc.co.uk/history/ww2peopleswar/stories/39/a4368639.shtml: accessed 24 January 2024).
[3] Ut supra.
[4] Ut supra.
[5] Ut supra.
[6] "Amsterdam 1930 HMHS - Hospital Carrier."
[7] "The sinking of the S.S. *Amsterdam* [Hospital ship] 1944."
[8] "Hospital Ships" (https://www.qaranc.co.uk/hospital-carrier-amsterdam.php: accessed 24 January 2024).
[9] Ibid.
[10] Ibid.

CHAPTER 20 NOTES:
[1] Bruhn, *We are Sinking, Send Help! The U.S. Navy's Tugs and Salvage Ships in the African, European, and Mediterranean Theaters in World War II*, 211.
[2] Ibid.
[3] Ibid, 211-212.

[4] Ibid, 212.
[5] Ibid.
[6] Ibid.
[7] Ibid, 213.
[8] Ibid.
[9] Naval Commander, Western Task Force, Preliminary Report of the Amphibious Invasion of Southern France, 1 October 1944.
[10] Ibid.
[11] Commander U.S. Eighth Fleet, Invasion of Southern France, 29 November 1944.
[12] Ibid.
[13] Ibid.
[14] Ibid.
[15] Ibid.
[16] Ibid.
[17] Ibid.
[18] Ibid.
[19] Ibid.
[20] Ibid.
[21] Ibid.
[22] "History of USS *Refuge* (AH 11)" Navy Department Office of the Chief of Naval Operations Division of Naval History (OP 29) Ship's Histories Section.
[23] Ibid.
[24] Ibid.
[25] Ibid.

CHAPTER 21 NOTES:

[1] Larson, *Army Hospital Ships in World War II*, 35.
[2] Ibid, 35-37.
[3] Ibid.
[4] Harris, *Troopships of World War II*, 331, 339-340.
[5] Ibid, 334, 347, 351
[6] "WW2 Hospital Ships" (https://www.med-dept.com/articles/ww2-hospital-ships/: accessed 24 February 2024)
[7] Larson, *Army Hospital Ships in World War II*, 37.
[8] Harris, *Troopships of World War II*, 340.
[9] Ibid, 331.
[10] Ibid, 334, 347.
[11] Ibid, 334.
[12] Ibid, 347.
[13] Ibid, 339.
[14] Ibid, 331.
[15] Ibid, 340.
[16] Ibid, 347.

CHAPTER 22 NOTES:

[1] "Gripsholm to Carry Exchange Prisoners," *The Morning News*, Wilmington, Delaware, 5 January 1945; "Trade Off: Exchanging German-Americans for POWs in WWII" by Jan Jarboe Russel (https://www.historynet.com/trade-off-exchanging-german-americans-for-pows-in-wwii/: accessed 14 February 2024).
[2] Headquarters Southern Line of Communications European Theater of Operations United States Army, Repatriation of Allied and German Prisoners of War and Exchange of Civilians, 8 January 1945.
[3] Ibid.
[4] Ibid.
[5] Ibid.
[6] Ibid.
[7] Ibid.
[8] Ibid.
[9] Ibid.
[10] Headquarters Southern Line of Communications European Theater of Operations United States Army, Repatriation of Allied and German Prisones of War and Exchange of Civilians; "The *Gripsholm* WWII Exchanges" (https://encyclopedia.densho.org/The_Gripsholm_WWII_Exchanges/: accessed 13 February 2024).
[11] A Full Life: An Autobiography of Stanley John Doughty 1921-1994 (published by his family).
[12] Ibid.
[13] Ibid.
[14] Ibid.
[15] Ibid.
[16] Ibid.
[17] Ibid.

CHAPTER 23 NOTES:

[1] "The Life and Times of Captain Morris O'Hara (Isaacs Harbour, NS), Master of Lady *Nelson* during the Second World War."
[2] Ibid.
[3] Ibid.
[4] "SS *Letitia*" (https://en.wikipedia.org/wiki/SS_Letitia: accessed 17 February 2024).
[5] "*Letitia* Canadian Hospital Ship" (http://www.forposterityssake.ca/Navy/LETITIA.htm: accessed 8 February 2024).
[6] Ibid.
[7] Canadian Hospital Ship *Lititia* (http://www.forposterityssake.ca/RCN-DOCS/LETITIA-01.htm: accessed 8 February 2024).
[8] Canadian Hospital Ship *Lititia* (http://www.forposterityssake.ca/RCN-DOCS/LETITIA-01.htm);

"Donaldson Dominion Duo: R.M.S. *Athenia* & *Letitia* Part Two"
(https://wantedonthevoyage.blogspot.com/2020/09/: accessed 8 February 2024).
[9] "Donaldson Dominion Duo: R.M.S. *Athenia* & *Letitia* Part Two."
[10] Ibid.
[11] Ibid.
[12] Ibid.
[13] Ibid.

APPENDIX D NOTE:

[1] "American Merchant Marine at War"
(http://www.usmm.org/hospital.html: accessed 24 February 2024).

Index

About the Author

Commander David D. Bruhn, U.S. Navy (Retired) served twenty-two years on active duty and two in the Naval Reserve, as both an enlisted man and as an officer, between 1977 and 2001.

He is a graduate of California State University, Chico, and has Masters degrees from the U.S. Naval Postgraduate School and U.S. Naval War College.

During his career, Bruhn served aboard six ships including command of the mine countermeasures ships USS *Gladiator* (MCM-11) and USS *Dextrous* (MCM-13) in the Persian Gulf. Ashore, he did two three-year tours in the Pentagon. During the first one, he was assigned to Secretary of the Navy and Chief of Naval Operation staffs as a budget analyst and resources planner. His final assignment was to the Secretary of Defense staff as executive assistant to a senior (SES 4) executive at the Ballistic Missile Defense Organization in Washington, D.C.

Following military service, he was a high school teacher and track coach for ten years, and remains an avid Track & Field fan. He lives in northern California with his wife Nancy and has two grown sons, David and Michael.

Bruhn has authored thirty-six books, which include 28 on naval history, 1 on army hospital ships, and 1 on shipboard engineering. In other subjects, there are 4 related to sports—*Toe the Mark*, *Stride Out*, *Distant Finish*, and *Beavers* about competitive running in the 1970s—1 about building a mahogany plywood camper for a lightweight truck, titled *Land Yacht Seaward*; and 1 about creating a British pub in a garage, titled *Stand Easy* (meaning take a break).